Flying the Cessna 210

The Secrets Unlocked

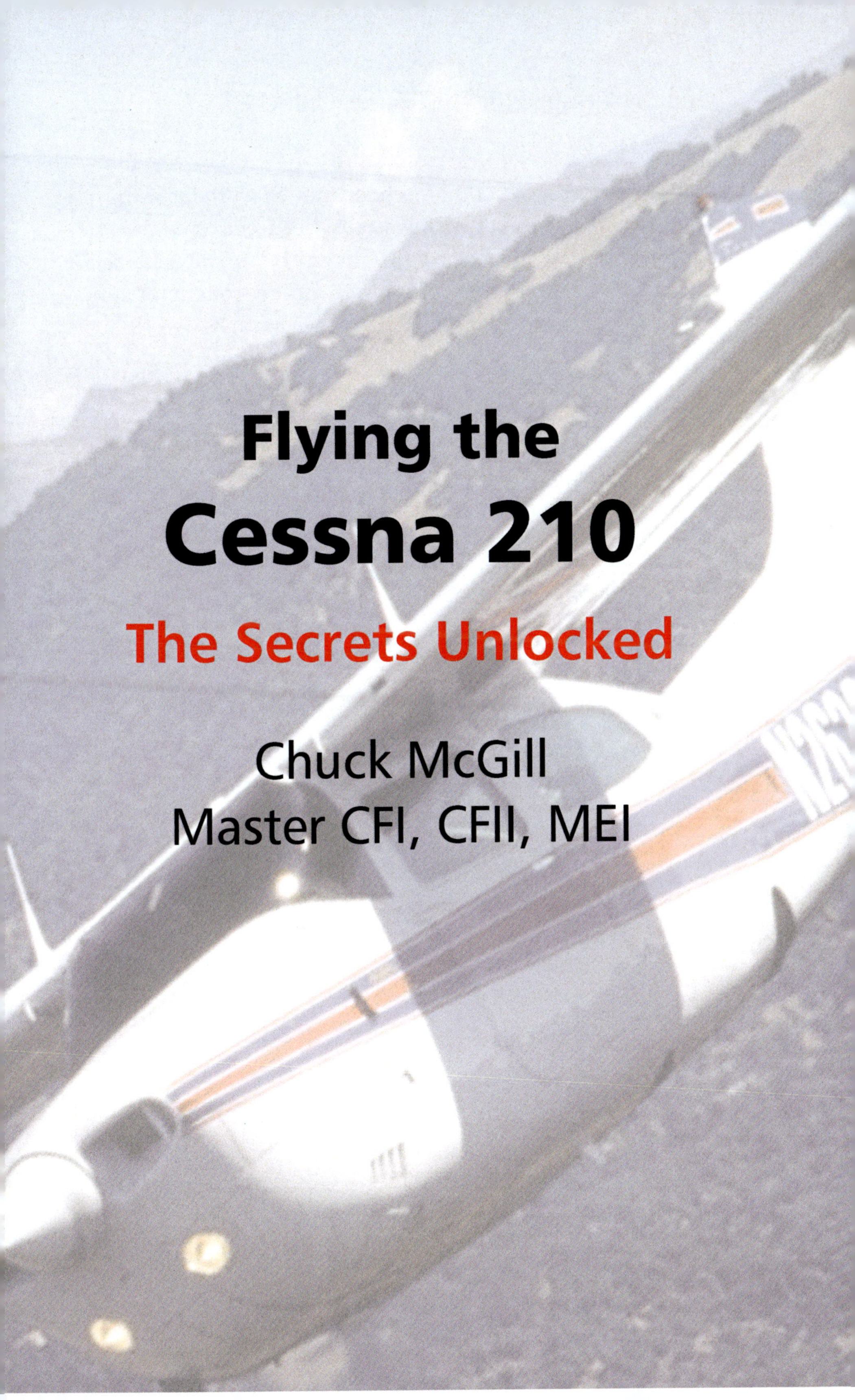

Flying the Cessna 210

The Secrets Unlocked

Chuck McGill
Master CFI, CFII, MEI

Flying the Cessna 210: The Secrets Unlocked

by Chuck McGill

Cover layout and design by Reya Kempley and Susan Cable.

Interior layout and design by Reya Kempley.

Edition ISBNs

Softcover	978-0-9884329-2-5
ePUB	978-0-9884329-0-1
Mobi	978-0-9884329-1-8
PDF	978-0-9884329-3-2

Library of Congress Control Number: 2012920327

Printed in the United States of America

Contents

Disclaimer

Nothing in this text supersedes any operational documents or procedures issued by the Federal Aviation Administration (FAA), the aircraft and avionics manufacturers, the Pilot Operating Handbook (POH), flight schools or the operators of the aircraft.

The author has made every effort in the preparation of this book to ensure the accuracy of the information. However, the information is sold without warranty either expressed or implied. The purpose of this book is to educate and entertain. The author and the publisher shall not be liable or responsible to any person or entity with respect to any loss or damage caused or alleged to be caused directly or indirectly by the information presented in this book. The information herein shall not be used in place of either common sense or the exercise of good judgment.

As stated in the Federal Aviation Regulations, Part 91, Section 91.3, Paragraph (a): The pilot in command of an aircraft is directly responsible for, and is the final authority as to, the operation of that aircraft.

© Paul Bowen

Introduction

Flying a Centurion can be a great pleasure because of the airplane's stability, speed, and excellent performance. Once you have mastered the Centurion, it can be a lot of fun to fly! You and your passengers will enjoy the speed and comfort of the airplane.

My experiences in owning, operating, and teaching in the Cessna 210 series of aircraft for over 25 years have been instrumental in the creation of this book. Many of my friends, relatives, and pilots who I have worked with as a 210 instructor have been encouraging me for years to put my thoughts and experiences to paper. This is that effort.

There are a multitude of reasons that I think the Centurion is one of the finest flying machines ever built. These traits and characteristics will be covered in various ways throughout this book. However, right now I would like to share with you that the airplane has excellent, if somewhat heavy, control response and has very predictable performance as you change configurations and power settings. Once you learn how to anticipate what the airplane performance will be when you change dynamics, the workload will diminish, and you will find the Centurion a joy to fly.

Although the Cessna 210 was last sold by Cessna as the 1986 year model, you will find thousands of Centurions flying all over the world in all types of environments. Centurions are flown mostly by private owners, but they are also flown by charter operations, freight companies, military organizations, law enforcement agencies, and a number of missionary organizations throughout the world.

The primary purpose of this book is to talk about how to fly the Centurion. It is not intended to be a historical tome or a guide about how to buy a Centurion or to maintain the aircraft, as those topics are thoroughly handled in other publications that are referenced in this book. This is about flying the Centurion in its various forms, and with the numerous modifications available to the airplane that expand, and in some cases, create limitations on the use of the airplane.

As with any endeavor, knowledge of flight and the aircraft are incredibly valuable tools, tools that can even surpass "natural talent" as a pilot. You are going to find a lot of information in this book of a different nature than you will find in your own aircraft's POH. But I am not going to bore you with lessons on aerodynamics, theory of flight, and all that, because I hope that those of you who are interested in flying a Centurion have already been schooled in those areas. This is not a flight primer — it is about flying this wonderful machine, and understanding its nuances and limitations as well as its capabilities.

Hours of study about the aircraft prior to flying it will enable you to transfer knowledge quickly into operational skill as you develop a feel for the airplane and learn to become the master of the aircraft.

This is a complex aircraft even in its most basic form. But when you start to modify it — and I would wager there aren't twenty original configuration 210s left in the world — you need to understand that you are changing the characteristics of the aircraft, and you are changing the manner in which it can, and in some cases, *must* be operated.

This book is intended to give you a thorough primer on the flight characteristics of the aircraft and the various manners in which it can be operated, depending on the installed equipment and modifications, if any, and the skill of the individual pilot.

I am not going to cover every model change or even every model year, but the bulk of the Centurions still flying are the 210L models and later (think of that as Centurions built in 1972 or later). Emphasis will be on the T210 and the P210, the aircraft I know best. A nice thing you will find here is that most of the systems and modifications discussed in this text or listed in the appendix will apply to almost all the 210s built, regardless of year or model.

A wide variety of performance information, including power settings, fuel flows, airspeeds, climb rates, configurations, and methods of controlling engine temperature extremes will be discussed. This information has been accumulated over a number of years flying a variety of Centurions. Even supposedly identical airplanes do not always perform the same at the exact same power settings and configurations. There are a few stock airframes out there, but almost all of the airplanes have different engines, different propellers or propeller efficiencies, different weight, different antennas, instrument error, wing modifications such as STOL kits and rigging of non-modified airplanes. All of these factors contribute to performance variations in the Centurion. So don't be surprised if you try a set of numbers shown in this book and they don't work out exactly the same for your airplane—there are just too many potential efficiency variables for them to match exactly (although they will match someone's airplane out there).

A caution: every FAA-approved modification to a Centurion comes with operating limitations. You, the pilot, need to read and understand the FAA-approved supplemental information for every modification to your airplane and every piece of installed equipment. This applies to the original optional equipment, such as that factory-installed autopilot, as it too has operating limitations.

This book is intended to provide useful supplemental information and a variety of operating techniques that may or may not apply to your aircraft. You are the only person who can know which tidbits of this information and which operating techniques can be applied to your specific airplane. Nothing herein suggests or implies that you ever deviate from the information and/or limitations in your own aircraft's Owner's Manual, POH, or AFM.

The thoughts and opinions expressed here come from over 4,000 flight hours in Cessna 210s, T210s, and P210s. I am not asking anyone to agree with my opinions or to use my techniques. I am using this medium to share my thoughts, experiences, and techniques with you, in the hope that they will expand your knowledge base and perhaps assist you in getting the most out of your Centurion.

When looking to explore in flight some of the things mentioned in this book, I suggest you first find a CFI who is qualified in the Centurion you are flying, be it a C210, a T210, or a P210. Experienced Cessna 210 CFIs are hard to find. Ask around, and once you find this instructor, interview him or her. This type of instructor isn't going to be the cheapest guy on the flight line, and if he or she is, really question their experience and knowledge level.

Any professional CFI will be interviewing you at the same time you are interviewing them. They will not be offended when you tell them what you are looking for and begin asking them to give you their experience level in these airplanes. Although you don't see a lot of lady 210 pilots, one of the best Centurion instructors I know is female — don't for a minute sell these ladies short. Historically, a significant portion of my students have been women, and I will tell you, there are many excellent female pilots who are also superior instructors.

Remember that you are hiring a CFI to teach you something, or to have a level of competence with your machine sufficient that you and he or she can experiment a bit; you are not hiring someone to ride along to build their flight time or enhance their resume. Your friend the Boeing 777 Captain or your neighborhood F-18 fighter pilot may be great in their jets, but nothing in those pilots' duties qualify them to teach you about flying a Centurion — find an experienced Centurion instructor. If you can't find one, contact the Cessna Pilots Association, AOPA, The Society of Aviation and Flight Educators (SAFE), or your insurance underwriter for a referral to an approved instructor. If you still come up dry, e-mail me through my website at www.SafeFlightIntl.com.

SAFE is an organization of professional flight educators, most of whom are Master Instructors, that can assist you in receiving high-quality ground and flight instruction. I am pleased to be a charter member of SAFE.

Mission Statement: The Society of Aviation and Flight Educators (SAFE) seeks to create a safer aviation environment through enhanced education. SAFE provides aviation educators with mentoring, support, and professional accreditation. By providing quality educational materials and other resources, we seek a reduction in aviation accidents, increased professionalism among aviation educators, and lifelong learning by everyone involved in aviation. Learn more at www.safepilots.org.

ALL NEW 210

C210

T210

P210N

1

The Airplane

C210, T210, P210

The Cessna 210, or Centurion, was first created in the late 1950s and entered the marketplace as a 1960 year model aircraft. A total of 9,304 C210s were built, according to Doug Miller, Archivist and Historian, Cessna Aircraft Company. At the time of this writing, there were 5,230 Cessna 210s listed in the FAA registration database; many others are still in use throughout the world.

1963 210D brochure photo

The aircraft was eventually available in three different versions: normally aspirated, turbocharged, and the turbocharged pressurized P210. Each aircraft offered different benefits related to speed, altitude, load-carrying capability, and passenger comfort.

All three versions were certified under Part 23 and are basically built around the same airframe; however, the P210 is structurally significantly different. The normally aspirated 210 and the T210 are almost exactly the same, except for different engines — and therefore some system differences which significantly alter performance and therefore require different operational procedures. Many of the basic characteristics of these aircraft are the same, including preflight and in-flight handling characteristics.

The original 210s had but four seats and a normally aspirated 260-hp engine. It was essentially a C182, wing struts included, with retractable gear.

1963 CESSNA 210

TOP SPEED at Sea Level	198 mph.
Cruise, 75% power at 7000 ft.	189 mph.
RATE OF CLIMB	1270 fpm.
BAGGAGE	120 lbs.
RANGE	
Cruise, 75% power at 7000 ft.	845 mi.
63.5 gal., no reserve	4.5 hr.
Optimum range @ 10,000 ft.	1530 mi.
80 gal., no reserve	11.2 hr.
SERVICE CEILING	20,300 ft.
GROSS WEIGHT	3000 lbs.
EMPTY WEIGHT (approx)	1780 lbs.
FUEL CAPACITY (standard)	65 gals.
(optional)	84 gals.

POWER: Continental 6-Cylinder, Fuel Injection IO-470-S Engine, 260 Rated HP at 2625 RPM

Four-place airplane

The first turbocharged Centurion sold was the 1966 T210F powered by a Continental TSIO-520C engine. Built-in oxygen was standard, and necessary, as the specifications claimed a 31,300-foot service ceiling.

In 1967, the Cessna 210G arrived with a laminar flow cantilever wing replacing the strut-braced wing, but it was still a four-plus-two aircraft, not yet a full-sized person six-seat aircraft.

1967 210G

The introduction of the 210K in 1970 brought us the full-sized six-seat aircraft and a 400-pound gross weight increase to 3,800 pounds. One of the changes that came with the K model was the tapered tubular steel landing gear legs with a six-inch wider stance. The gear legs look fragile, but they aren't.

The 210L arrived in 1972 with a 28-volt electrical system, and in 1975 the three-bladed propeller became standard. Flap gap seals in 1976 enhanced climb performance and a key instrument change occurred: the airspeed indicator was changed to KNOTS from MPH.

The M model of 1977 ushered in the 310-hp TSIO-520R engine for the turbocharged airplanes, which reduced takeoff distances and provided increased climb rates. The engine has a five-minute limit on the 310 hp, which then must be reduced to 285 hp maximum continuous power.

The Pressurized 210 was introduced in 1978 as the P210N. It was powered by a Continental TSIO-520P putting out 310 hp at takeoff, and like the other turbocharged 210s, was limited to 285 hp after five minutes. Records indicated that a total of 851 P210s had been built when production ceased.

Gear mirror

In 1979, the main landing gear doors were removed from the Centurions. Many of the earlier Centurions had their main gear doors removed by owners with a modification called the Uvalde Gear Door Mod. My own aircraft has that modification, which I had performed immediately following an IFR flight where I could not get a green gear-down light. Because I didn't have gear mirrors at that time, I couldn't see the position of my wheels, which made it a most unpleasant event. The value of gear mirrors was instantly implanted into my brain on that flight.

In 1982, there was a major change in the fuel system. Cessna accomplished this by removing the dual fuel reservoir tanks and installing a single reservoir tank. This allowed the pilot to select Left, Right, or Both fuel tanks in flight. Obviously, being able to select both fuel tanks reduced the pilot's workload. Other fuel system changes included separate fuel vapor return lines to return the vapor to each fuel tank.

The economy was in a spin during the mid-1980s and sales of new aircraft skidded to a halt. This gave Cessna time to redesign the 210 for the 1985 release of the R model, which was sold new as the 1985 and 1986 models. There were many changes in the '85 model; one of the most significant being enlarging and strengthening the tail of the aircraft in a variety of ways, including making the horizontal stabilizer a full three feet wider! The wing span was increased to 38 feet, 10 inches and the gear extension speed and flap speed were raised. The bottom line is: the 1985/1986 210R is a very different airplane than its predecessors and they are highly prized by their owners. Cessna delivered their last three Centurions in 1987 as 1986-year models.

T210 at Tsuniah Lake, Canada

By any definition, the Centurion aircraft are complex airplanes. Original systems are in some cases very complex and perhaps not thoroughly explained in the Owner's Manual, POH, or AFM. As time has passed and technology has changed, we now see extensively updated avionics and various other modifications made to each aircraft that complicate the situation even more.

This is a *checklist airplane*! It is very important to your safety that you ensure the preflight and flight operations checklists you use for this airplane are representative of the actual aircraft you are operating. One checklist *will not fit all*! Seemingly identical aircraft are probably not, and a switch in the wrong position or a circuit breaker located in a different position can cause harm to the aircraft or you and your passengers.

In my opinion, the Centurion is a very safe, comfortable, and easy aircraft to fly if you have been well trained and are proficient in it. Unfortunately, it is an aircraft that has a less-than-admirable safety record, with over 30 reported in-flight airframe failures. This aircraft is capable of staying together at 5.7 Gs, 150% of its 3.8-G rating, so you may wonder: why the in-flight airframe failures? I will get to that later.

Fuel management has also historically been a problem and continues to be a leading cause of Centurion accidents, despite it being a widely known issue with the aircraft.

NTSB Identification: CEN11LA402
Accident occurred Saturday, June 18, 2011 in Albia, IA
Probable Cause Approval Date: 06/14/2012
Aircraft: CESSNA P210R, registration: N900SR
Injuries: 1 Serious

The pilot stated that the airplane was fully fueled before takeoff. While en route, he requested a destination change due to deteriorating weather along his route of flight. During the descent, about 15,000 feet mean sea level, the engine lost power. The pilot's efforts to restore engine power were not successful, and he diverted to the nearest airport. The airplane descended below the clouds about 1,000 feet above ground level; however, the airplane was not in a position to land on the runway. The airplane ultimately lost too much airspeed and altitude as the pilot attempted to configure and align the airplane for a landing on the runway. He subsequently landed in a grass area about 100 feet southwest of the runway, nearly perpendicular to the runway orientation.

A post-accident examination did not reveal any anomalies consistent with a pre-impact failure or malfunction. No usable fuel was observed in the airplane fuel tanks at the time of the examination. The fuel tank caps were

> securely installed, and no evidence of fuel staining or siphoning was observed on the airframe. The pilot noted that the fuel quantity indications before the loss of engine power were near empty and about 10 gallons for the right and left tanks, respectively. He added that the fuel gauges had never worked properly. The pilot stated that he did not visually check the amount of fuel on board during the preflight inspection. He commented that normal airplane endurance was about 5 ½ hours and the accident flight was about 3 ½ hours in duration. The airplane was not in compliance with a Federal Aviation Administration Airworthiness Directive requiring a placard denoting that the fuel level must be rechecked 2 minutes after fueling to ensure full fuel tank capacity. In addition, the airplane Pilot's Operating Handbook required a visual inspection of the fuel level as part of the preflight inspection.
>
> The National Transportation Safety Board determines the probable cause(s) of this accident as follows:
>
> The pilot's inadequate preflight inspection, in which he did not detect that less fuel was on board the airplane than planned, which precipitated a loss of engine power due to fuel exhaustion during initial descent.

The AOPA Air Safety Foundation has published a Safety Review of the Centurion; it is filled with valuable information and accident analysis for the owner or operator of a 210. I highly recommend you purchase a copy.

Gear-up landing on a beach

If you would like to do some interesting research on the Centurions, you can go to the www.FAA.gov website and pull up the Type Certificate Data Sheet on the entire 210 series. The Type Certificate Data Sheet details are contained in a single 45-page PDF file. The Type Certificate Data Sheet file is a treasure trove of information for those who want to know exact details, especially limitations, about a specific make and model. The information is detailed right down to the serial number of the airplane.

If you would like to study an easy-to-read, in-depth and detailed history of the Cessna 210's model year changes, I suggest you purchase *John Frank's Cessna 210 Centurion Buyers Guide*, published by the Cessna Pilots Association in Santa Maria, CA. John has spent what seems like a lifetime working with Cessna 210s and is probably the most knowledgeable person around when it comes to maintenance and aircraft systems of the 210.

Cessna Pilots Association™

2

The Wow Factor

Speed, Altitude, Weight-Carrying Capability, Comfort, Stability, Short Field Capability

Speed: Fast for a Single-Engine Six-Seat Airplane

The normally aspirated Cessna 210 is very fast for a six-seat single-engine airplane at up to 174 KTAS; its only competition for speed in its class is the Beech Bonanza. An engine modification that puts a Continental IO-550 in the C210 gives a substantial increase in climb rate and TAS.

The T210 is the fastest of the Centurions with up to 202 KTAS at Flight Level 240, with typical mid-teen altitude speeds in the 170-180 KTAS range.

The P210 can cruise at up to 196 KTAS at optimum altitude, but routinely delivers much lower speeds, such as 160-165 KTAS at normal cruise power settings. Some owners drive them hard and claim 175 KTAS in the upper teens.

Altitude: These Airplanes Can Fly High, Over Most of the Weather!

The normally aspirated Cessna 210 has a very high service ceiling for a non-turbocharged airplane, ranging from 15,500 feet to 17,300 feet, except for the '64 and '65 year models, which have service ceilings of 21,000 feet. Few of these aircraft have built-in oxygen, and most are flown well below their service ceilings, typically in the seven- to twelve-thousand-foot altitude range.

Clear, smooth air above the clouds

The T210 isn't only fast, it can fly between 27,000 feet and 31,300 feet depending on the model — although routine flight above FL250 is flown by only a few of the pilots I have met. These airplanes fly really well and deliver excellent speed and economy in the mid-teen altitudes. With about 80 percent of cloud tops below 18,000 feet, this is a valuable ability and normally includes a smooth ride!

Overhead oxygen connections are nicely positioned for ease of use

The P210N has a maximum operating altitude, limited by cabin pressurization, of 23,000 feet, with the P210R having a 25,000-foot limit. They are typically flown in the mid to upper teens unless higher altitudes are needed for cloud clearance, as they are not sprightly in climb performance.

Weight-Carrying Capability: Huge

Normally Aspirated: An enormous useful load of up to 1,600 pounds can be found in the POH, but typical actual useful loads run about 1,500 pounds and above, which is still a large useful load for a single-engine airplane.

Turbocharged Centurions: Useful loads of 1,400-1,600 pounds are routinely found, although the 1985-year model AFM claims an amazing 1,798 pounds on a lightly equipped airplane.

Even the bicycles are going!

Pressurized Centurions: The useful load falls off from its siblings, because of the heavier airframe and additional systems, to 1,200-1,300 pounds typically. Although some are higher than 1,300 pounds, some are actually below 1,200 pounds because they are carrying the weight of known ice certification, air conditioning, and auxiliary fuel tanks.

Cabin: Comfortable, Roomy, and Versatile

Wide, comfortable cabin

Three rows of seats with one center seat removed

P210 interior

The Centurion is a true six-adult-with-baggage airplane. The cabin is wide and the ceiling high. The folks sitting in the aft bench seat need to be fairly short to ensure comfortable leg room and that their heads don't touch the ceiling, but it works. You want the lightest-weight individuals in the back seat for proper Weight and Balance, and that is usually the shortest folks, so it all works out. When I was using my aircraft for a time in a business venture, I would carry six adults fairly often — but the stage lengths were less than an hour, so no one whined excessively. The back seat is ideal for children or, as in our case, dogs.

Middle and rear seats shown in a P210

The front seats are adjustable and the middle seat backs can be adjusted for those who like to lean back when seated. Pitch and yaw moments in the front and center rows of seats are minimal. My

Pilot and copilot seats

"The Queen's seat" is shown; the other center seat has been removed

Rear seat removed for a three-passenger configuration

wife and I refer to the center seat behind the co-pilot as "the Queen's seat" because it is so comfortable.

The cabin is well ventilated for cooling, and a retrofit kit is available for changing out the old plastic vent vanes for airline-style vents. That retrofit is well worth the money and a must-have if you are redoing the interior. Heat output is a bit weak if you are sitting in the aft seat; a small comforter is carried in our airplane during cold weather for the back seat passengers. There are noticeable

Airline-style air vents are an aftermarket modification

Airline-style air vents for all passengers

pitch and yaw moments when riding in the rear most seat, but it's nothing unmanageable for the passengers—unless the pilot is not using the rudder pedals.

Cabin noise levels are fairly low in the second and third seat rows, and those passengers can chat with each other during flight without using the intercom.

Stability In Flight: Solid

The Centurion is a very stable aircraft for its size and weight, which helps make it an excellent IFR platform. Both elevator and rudder trim are standard in the airplane. The ailerons have ground-adjustable trim tabs. If you trim the aircraft properly during flight, you may think the autopilot is on. If you don't trim it properly, you will be doing a lot of pushing and pulling to try to hold altitude and course. The controls are fairly stiff, especially so in the P210 because they are routed through the pressure vessel. This is an airplane that requires you to use the trim to relieve control pressures, or you can get your daily workout by fighting the airplane. The Centurion also needs a lot of rudder input as you fly it through the various flight regimes. Right rudder trim on takeoff and for climb is absolutely necessary, especially when you are climbing to the flight levels. When turning, as with most aircraft, if you want a nice coordinated turn, you need to lead with the rudder and follow with the hands.

Short-Field Capability with a STOL Modification: Amazing!

Operating out of short fields in an unmodified Centurion will yield some fairly short takeoff runs when the engine and prop are doing their job properly and the pilot is proficient. Depending on the temperature and altitude at which you are operating, you can be airborne in less than a 1,200-foot ground roll in many of the Centurions, even at a gross weight of 4,000 pounds. However, if you wish to always have a lower takeoff speed and a lower stall speed with reduced takeoff and landing distances, adding a Short-Field Takeoff and Landing (STOL) modification is an excellent and inexpensive option.

STOL Cessna 210 at an Idaho backcountry strip

Short-field takeoff and landing modifications for this airplane can turn it into an even more amazing machine. Before operating in STOL mode, the pilot needs to have taken STOL training from a qualified instructor and be on top of his game. When the pilot has received good STOL training and he is on his game, the modified 210's performance is incredible. Takeoff speeds from 46-52 knots in a 3,800-pound airplane are pretty impressive and you need less than 1,100 feet at gross weight to be airborne. Approach speeds as low as 47 KIAS can yield very short landing distances: less than 1,000 feet over a 50-foot obstacle. But again, the pilot has to be well-trained and at his best to achieve these results.

You don't have to operate the aircraft in STOL mode to benefit from the STOL modification. The STOL modification still helps you during routine operations because it gets you airborne sooner and allows you to land at a slower speed (because the stall speed is lower) if you fly it properly. Further, the STOL modification takes away the tail waggle found during low speed approaches in the pre-1985 airplanes. The lower stall speed makes for easier and softer touchdowns while saving wear and tear on the tires in both takeoffs and landings.

STOL modifications will not be found on known-ice-certified airplanes regardless of whether they are using boots or a TKS anti-ice system. However, vortex generators (VGs) can be found on them, which do assist in lowering the stall speed and removing the little tail waggle.

Two different types of STOL modifications can be found on Centurions. A very simple but efficient design such as the Horton STOL kit incorporates a leading edge cuff and stall fences. A more complex Robertson (now R/STOL) STOL yields about the same results as the Horton but incorporates drooping ailerons as well as other high-lift devices. You will find more information about the STOL modifications and their related companies in Appendix C.

Owner's Manual

Pilot's Operating Handbook (POH)

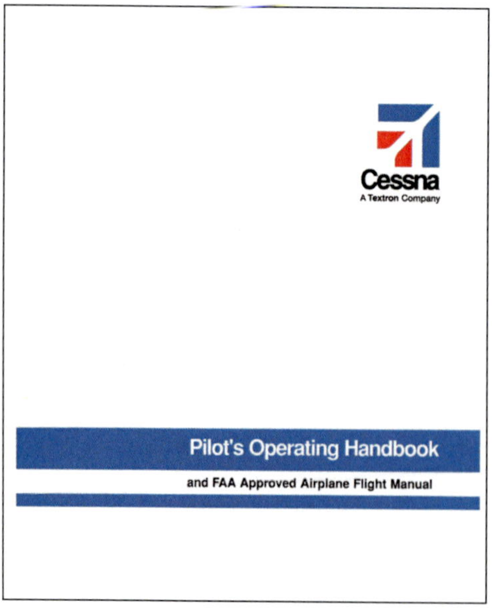

Pilot's Operating Handbook (POH) and Approved Flight Manual (AFM)

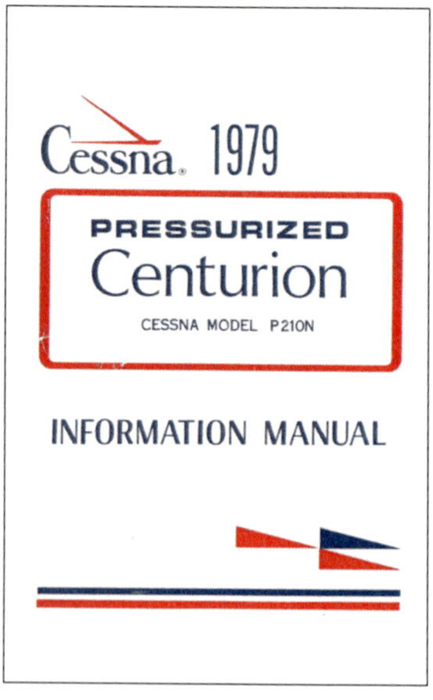

Pilot Information Manual (PIM)

3

Owner's Manual, POH, or AFM

Which one do I have? What's in it?

Owner's Manuals were published pre-1976 for the Cessna 210s. They were a publication of the times, before the general aviation industry matured and more standardized data was required to be presented to pilots. These were the times when ARROW (airworthiness certificate, registration, radio station license, Owner's Handbook or Manual, and Weight and Balance) was a real necessity. These thin manuals were basic at best and in some cases had confusing information in them. They lacked detailed performance and systems information and rarely had any information on optional equipment, such as radios. All of the information needed to operate the aircraft was not in a single document. Placards needed for operation were found on the instrument panel and various other locations on the aircraft. The Weight and Balance worksheet was floating around somewhere — but not in the Owner's Manual, because there was no place for it, unless you somehow secured it to that flimsy manual.

A cooperative agreement between the General Aviation Manufacturers Association (GAMA) and the FAA led to standardized POHs by all U.S. aircraft manufacturers in 1976. The POH is a more complete document than an Owner's Manual, with detailed performance information, aircraft systems information, and supplemental data on the manufacturer's optional equipment.

In 1978, the FAA instituted a procedure whereby an FAA representative signed every POH at the factory. These signed POHs were then called Approved Flight Manuals, or AFMs. They look and feel like a POH and most pilots call them a POH.

Cessna continues to publish a lower cost "Information Manual" on the later year model aircraft that is basically a non-approved POH or AFM. They do not contain aircraft-specific information (actual Weight and Balance, flight manual supplements for optionally installed equipment, etc.) and therefore cannot be used as the required POH or AFM in the aircraft. However, they are an excellent source of original information as of the date of their publication and I have used a number of them for reference in preparing this book.

Information Manual

Many of the Information Manuals have the most current POH/AFM revisions; I found a revision in one of them that was not in my AFM for one of my airplanes. I therefore had to query Cessna and purchase the revision from them. It seems the mailing address on my aircraft registration did not have my complete delivery address and the revision Cessna had sent out was returned to them as undeliverable. Another lesson learned.

If you are renting an airplane and don't have regular access to the Owner's Manual, POH, or AFM, I encourage you to purchase an Owner's Manual or Information Manual for that aircraft. Be sure to add the necessary supplements and Weight and Balance for the aircraft you operate and you will have a duplicate POH/AFM.

Regardless of the make or model of aircraft you are flying, the POH/AFM always presents the information in a standard format as shown here:

SECTION 1	GENERAL. A brief overview of the aircraft.
SECTION 2	LIMITATIONS. Detailed limitations of the basic aircraft *not* including optionally installed equipment, avionics, etc.
SECTION 3	EMERGENCY PROCEDURES. Self-explanatory.
SECTION 4	NORMAL PROCEDURES. Preflight to landing—but does *not* include optionally installed equipment.
SECTION 5	PERFORMANCE. Charts on everything about flying the airplane.
SECTION 6	WEIGHT & BALANCE/EQUIPMENT LIST. Self-explanatory.
SECTION 7	AIRPLANE & SYSTEMS DESCRIPTIONS. All the details of what the systems are and how they operate. Does *not* include optional installed equipment, avionics, etc.
SECTION 8	AIRPLANE HANDLING, SERVICE, AND MAINTENANCE. Required inspections, ground handling, servicing systems, care and cleaning.
SECTION 9	SUPPLEMENTS (Optional Systems Descriptions and Operating Procedures). This is a listing of all optional equipment that was available from the factory when the aircraft was manufactured. Information on systems that were not available from the factory will only be found here if the owner put it in the POH/AFM as he should have (see below).

This is very important. The POH Section 9—Supplements, provides required Optional Systems Description and Operating Procedures. Whenever you or a prior owner adds new equipment to the aircraft, the installing facility is supposed to provide the FAA-approved Flight Manual Supplement for that item. The owner is supposed to place the new Flight Manual Supplement in Section 9 of the POH or AFM. Obviously, you should remove supplements for equipment not installed, or that you have removed, and create a new index for Section 9 so you will be able to quickly find system information in the event of an emergency. Please note that the Emergency Procedures section does not contain emergency procedures for optionally installed equipment! This information is only found in the Flight Manual Supplement.

SECTION 9
SUPPLEMENTS
(Optional Systems Description & Operating Procedures)

TABLE OF CONTENTS

Section 9 Table of Contents: Before

CESSNA
MODEL T201L

Section 9
SUPPLEMENTS

SECTION 9
SUPPLEMENTS
(Optional Systems Descriptions & Operating Procedures)

TABLE OF CONTENTS

Note:

The GTS800 Traffic System displays on the GTN750
The GTX330 Traffic System displays on the GNS430W
When using Transponder #2 TIS traffic will not be displayed

9-1

N262CM October 2011

Section 9 Table of Contents: After

In addition to placing the Flight Manual Supplement into Section 9 of the POH/AFM, the pilot needs to read it! This is where the operating procedures *and limitations* of the newly installed equipment are found. Failing to read the Supplement means it is highly unlikely you will know how to properly operate the system, much less know the limitations of the system. This may sound simple, but in an airplane with a number of complex systems that have been added over the years, it can become a serious issue if you don't operate the systems correctly and integrate their operation properly. An example might be turning the autopilot on at an indicated airspeed below its operating speed—and then wondering why the autopilot won't work. Or perhaps extending speed brakes in icing conditions only to find out you can't retract them, as you unhappily observe a dramatically increasing rate of descent.

Speed brake in deployed position

Many of the Centurions flying today have had extensive avionics modifications. Complete "glass panels" are becoming the norm. While this is mostly a good thing, *dangers lurk behind the scenes* when all the systems are changed and the information in the POH/AFM no longer represents the actual aircraft systems.

Let's take a typical Garmin G500 installation in a P210 as an example of what might occur during a major avionics upgrade. When the P210 was originally delivered, it had at least one, if not two, vacuum pumps. If it had a Cessna autopilot, the autopilot depended on input from the vacuum-powered attitude indicator to operate. It probably had a single 90-Amp alternator. The pitot static system provided input to mechanical airspeed indicators, altimeters and vertical speed indicators. The engine instruments were mechanical or electromechanical.

1982 P210 Factory Demonstrator instrument panel — before renovation

A detailed diagram of the electrical system was presented in Section 7 — Systems. A photograph of the instrument panel was shown in Section 7, along with detailed descriptions of every knob, switch, button, control device, or indicator.

Along with the installation of the G500 came a new engine monitoring system; for example, a JPI930. Along with new glass flight instruments

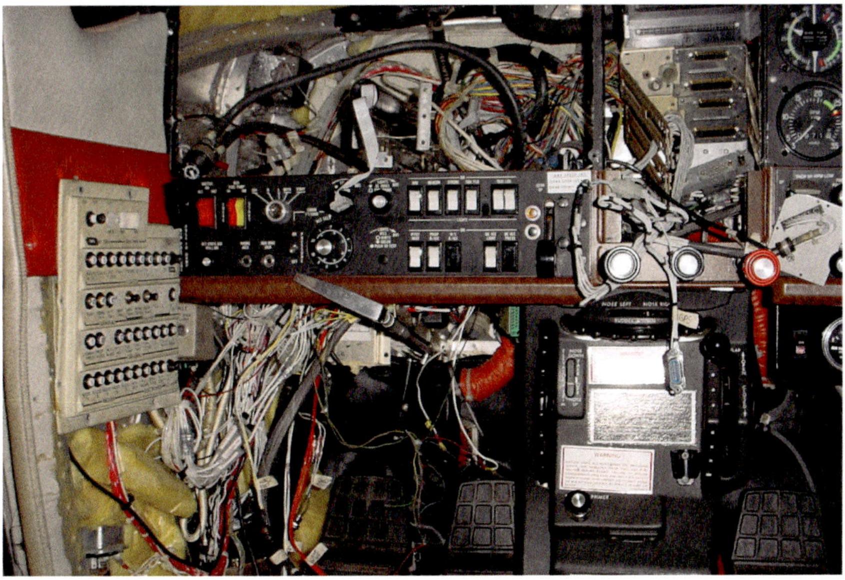

Major avionics renovation in process

Garmin G500 panel installation completed

came the *removal* of all the original engine gauges, including fuel gauges, oil temperature and pressure, propeller rpm, etc. The turn coordinator is now gone. The VSI is gone. The compass remains, but nothing on the panel except for the FAA-required mechanical backup attitude indicator, airspeed indicator, and altimeter are retained. Everything else is electrical. To provide power to the portable GPS and the iPad Electronic Flight Book (EFB), two or even three new electrical outlets were added. Of course, a second alternator was also added for "just in case."

Now the electrical system bears no resemblance to the schematics in Section 7. The vacuum system is gone, so that portion of Section 7 is invalidated, as well as anything in Section 9 regarding vacuum. You have no idea how the system is wired or what the electrical loads of the various systems are.

When this type of extensive modification occurs, there are more and different circuit breakers, perhaps more electrical buses, and in many

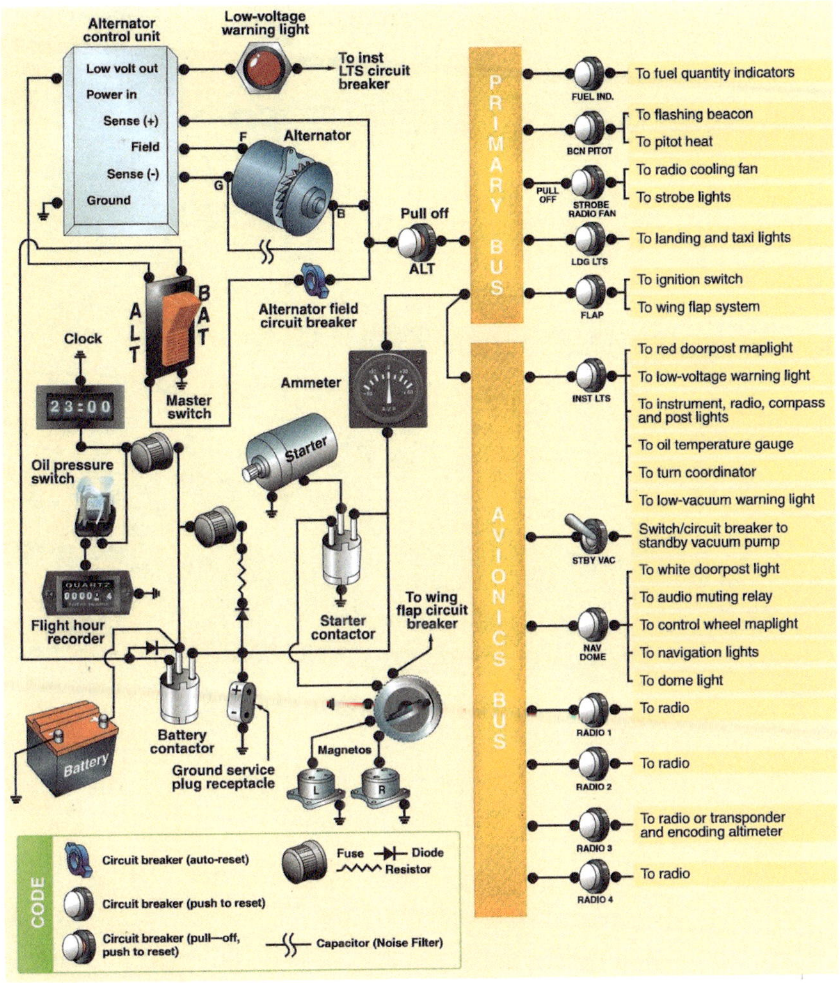

Figure 6-34. *Electrical system schematic.*

Electrical system schematic

cases more circuit breaker panels. The result is that the cockpit representation showing us the location of all the switches and dials in Section 7 of the POH/AFM no longer represents the aircraft.

If and when a "redo" of the panel is done, we need to ensure the shop that does the work provides us with current electrical schematics and a new Electrical Load Analysis. (The original Electrical Load Analysis

New circuit breaker panel to accommodate avionics changes

is found in the Service Manual for the aircraft, which few owners have or have ever seen. It is normally only found in maintenance shops.)

Once the installation shop has provided you with the necessary schematics, which will allow you and a service facility to be able understand the new systems and how they are integrated, take a photo of your panel and all the circuit breaker panels, etc. Print out and label the photos so that you and any other pilot can look at it and gain an understanding of the "new" panel and its related idiosyncrasies. Also, any maintenance shop that is going to work on the airplane is going to need to know this information.

You need to create or have created new portions of Section 7 for your POH/AFM that reflect the aircraft as it is today, so the aircraft can be operated, maintained, and repaired properly.

Of course, now comes the question of preflight, flight, and *emergency* checklists, which will be covered in another chapter.

4

Weight and Balance

It can carry a load!

Every pilot wants to know, "What is the useful load?" It seems we all want to fill up the seats, fill up the fuel tanks, and fill up the baggage compartment. If you try that with most four-place airplanes, they suddenly become two-place airplanes. When most six-place airplanes are fully fueled and loaded, they are at best four-place airplanes. Well, the Cessna 210 can usually take all six people, full fuel, and the ski gear as well, unless it is heavily equipped.

You can find many Cessna 210s with useful loads of over 1,400 pounds, and T210s with useful loads of up to 1,600 pounds, depending on the year model and how much optional equipment has been added on.

Before moving on to specifics by make and year model, let's take a look at things that materially reduce the useful load on these airplanes. We don't have to look far, as just three items can account for a lot of weight. Those three items are:

- Autopilot with an HSI (50.1 to 61.7 pounds depending on system type)
- Known icing certification (37-50 pounds depending on the year model)
- Air conditioning (65+ pounds)

Most of these airplanes have an autopilot, and if you are flying real IFR I think you shouldn't be without one. The need for known ice certification depends on where you live and what kind of weather you are inclined to fly in. Air conditioning is a nice item to have in the Cessna 210 and the T210, but pretty much a required item in the P210 due to the absence of fresh air when pressurized.

If you have a 210 with an old Cessna autopilot, known ice certification, and air conditioning, those three options will eat up roughly 150 or more pounds — the weight of a light passenger. Changing out the autopilot to a newer digital type, such as the S-TEC 55X, and upgrading your avionics to digital radios and GPSs will pick up a nice bit of useful load. If you upgrade the flight instruments to the Aspen Pro Series or the Garmin 500 series, you will pick up even more useful load as the old heavy mechanical instruments and their accompanying hardware is replaced by much lighter electronic devices.

The Cessna 210 had a 3,800-pound gross weight starting in 1970 (K model); a lightly loaded 210 could conceivably have a 1,620-pound useful load. It wouldn't have much optional equipment, so you don't really see many 210s with a 1,600-pound useful load, but many are in the 1,500-pound range. The gross weight remained at 3,800 pounds until 1985 with the R model, when it went up slightly to 3,850 pounds. The few R models that were produced were well-equipped, and therefore resulted in no effective increase in useful load despite the slightly increased gross weight.

The T210 can have a huge useful load, depending on the year model and the optional equipment installed. Starting in 1970 with the K model and ending with the M model in 1978, they had a 3,800-pound gross weight. Lightly equipped T210s of these years can be seen with useful loads from the high 1,300s to the low 1,400s. (My personal T210 has a 3,800-pound gross weight, but since all of my avionics have been updated, I have a very satisfactory 1,400-pound useful load.)

The 1979 T210N brought a 200-pound gross weight increase that really helped even the most heavily equipped T210s. With great avionics, known ice certification, and air conditioning, you still could have a 1,400-pound or above useful load.

The T210N is the real heavy hauler of the pack. If gross weight/useful load are key factors in your flight profile, then a '79 or later T210N may be right for you.

The P210 is a wonderful cross-country flying machine. Older pilots like the pressurization, as do those pilots with young children or dogs that travel in the airplane regularly. The first P210 out of the factory in 1978 had a gross weight of 4,000 pounds. *But* the P210 is heavier than the C210 or the T210, and they are usually well equipped. The P210 basic weight is higher that its stablemates primarily due to the additional weight of the pressure vessel and the pressurization system components. Some P210s are still carrying around nearly 30 pounds of radar-related components in addition to a flight director system/autopilot, known ice certification, and air conditioning. The P210 is usually found with a useful load in the mid-1,300 pound range, although lightly equipped P210s can be found with a useful load of 1,400 pounds or more. This makes the P210 something less than a real six-passenger airplane.

All but the last 39 P210s were N models. In 1985, Cessna introduced the P210R which had a gross weight of 4,100 pounds. It was not much of an increase over the N model.

We all know that the airplane's performance is seriously impacted by weight. But how much does your Cessna 210 really weigh? This is a serious question. What is the "actual" weight of your airplane? Virtually every 210 flying today has been modified in some way. There is a saying that "airplanes are like people, they gain weight as they age."

Let's look at a 1979 Cessna 210 aircraft. It doesn't matter if it is a C210, T210, or P210; the odds are it has been modified. Perhaps it has a different interior or just new carpet. Perhaps it has been repainted or worse yet, over-painted, so that it is now carrying double the weight of paint than should be on the aircraft.

The avionics have probably been changed as Loran, ADFs, and their related antennas went away. Were the antennas taken off and coaxial cables removed? Some of these modifications and changes have been recorded in the Weight and Balance data, and some have not. The Weight and Balance worksheet that you find in Section 6 of your POH/AFM is a calculated Weight and Balance; how many times has

Cessna 210 chemically stripped for painting

it changed? If a mistake was made early on, it has probably been carried forward.

The only way you are going to know the actual weight of your airplane is to have it weighed. Yes, pay the FBO to put it on the scales and find out how much it really weighs.

Now, the FBO is going to empty the airplane to weigh it. All of your assorted "stuff" that you have in the airplane has to come out for it to be weighed. All of the old charts that are under the seats, the extra headsets and chocks, oil, step ladders, funnels, the flashlights and extra microphones — everything comes out!

The good news about that is now you can throw away all of your accumulated junk and five-year-old candy bars. Once removed from the airplane, put all of your additional flight equipment that you have chosen to keep in a box and weigh it! Once you have determined your "standard additional weight," you can you put everything back in the airplane and add that weight to the actual weight of the airplane before computing your Weight and Balance. My standard additional weight, including my flight bag, is a whopping fifty-two pounds, primarily because of the thirty-five pounds of survival equipment I carry.

CESSNA AIRCRAFT COMPANY

WICHITA, KANSAS

Weight & Balance and Installed Equipment Data

Revised

CONTROL NUMBER	DATE	MODEL	REGISTRATION NUMBER	CESSNA SERIAL NUMBER
237035	02/07/83	PRESS. CENTURION II	N6586W	P21000800

ITEM		WEIGHT	ARM	MOMENT
STANDARD AIRPLANE (EMPTY,DRY)	COMPUTED	2378.6	40.6	96571
INCLUDING ALL REQUIRED AND STANDARD EQUIPMENT ITEMS				
UNUSABLE FUEL 3.0 GALLONS		18.0	38.0	684
FULL OIL 11.0 QUARTS		20.6	11.8-	243-
STANDARD EMPTY WEIGHT		2417.2	40.1	97012
OPTIONAL EQUIPMENT REPLACING OR IN ADDITION TO				
REQUIRED AND STANDARD EQUIPMENT ITEMS				
-W-ALLOVER VESTAL WHITE	NEGL.			
-A-CHOCOLATE BROWN-MAJOR	NEGL.			
-G-MOCHA BROWN-ACCENT	NEGL.			
-F-SUNBURST YELLOW-THIRD COLOR	NEGL.			
CARPET BROWN-MONACO	NEGL.			
SEAT INSERTS HANOVER BROWN FABRIC	NEGL.			
SEAT BELTS BROWN	NEGL.			
SEAT TRIM BROWN REEVANE	NEGL.			
LOWER SIDEWALLS BROWN VINYL	NEGL.			
UPPER SIDEWALLS BROWN VINYL	NEGL.			
WOODGRAIN UPPER INSTRUMENT PANEL	NEGL.			
BASIC AVIONICS KIT		6.3	56.5	356
400 TRANSPONDER RT459A HIGH ALTITUDE		3.8	17.1	65
INDICATOR, TRUE AIRSPEED (EXCH)		.1	18.0	2
DETECTORS, NAVIGATION LIGHTS	NEGL.			
FLIGHT CONTROLS, DUAL		7.3	14.1	103
GROUND SERVICE PLUG RECEPTACLE		1.9	3.9-	7-
HEATING SYSTEM, STALL WARNING & PITOT (EXCH.)		.2	26.3	5
LIGHT, BEACON,OMNI-FLASH		1.6	226.0	362
LIGHT, COURTESY (L.H. WING ONLY)		.3	51.4	15
LIGHTS, INSTRUMENT POST		.5	19.5	10
LOCATOR BEACON		3.5	159.0	557
STATIC KIT, ANTI-PRECIPITATION		.4	154.8	62
SEAT, CO-PILOT'S, ARTIC RECLINE & VERT ADJ.(EXCH)		5.8	25.1	146
COM ANTENNA AND NAV COUPLER		1.1	41.2	45
400 GLIDESLOPE R443B 40 CH. INCL. VOR/ILS (EXCH)		3.9	78.4	306
400 MARKER BEACON (R402A)		2.5	88.0	220
ALTITUDE ENCODER, BLIND -1000 TO 30,000 FT. RANGE		1.6	16.6	27
AIR CONDITIONING, CABIN		80.2	62.3	4996
ALTERNATOR, 28-VOLT (95 AMP) (EXCH)		4.8	5.0-	24-
CLOCK, DIGITAL QUARTZ READOUT (EXCH)	NEGL.			
FIRE EXTINGUISHER, HAND TYPE		3.0	35.0	105
APPROACH PLATE HOLDER		.1	22.0	2
LIGHT, ICE DETECTION, LH		.6	9.6	6
DEICE, PARTIAL PLUMBING (WING AND STABILIZERS)		4.2	48.9	205
VACUUM SYSTEM DUAL PUMP (EXCH)		4.3	1.1	5
HEADREST (SET OF 2-FRONT)		1.8	48.0	86
LIGHTING, ELECTROLUMINESCENT		2.3	18.0	41

Original Weight and Balance with modification (p. 1)

CESSNA AIRCRAFT COMPANY

WICHITA, KANSAS

Weight & Balance and Installed Equipment Data

Revised

CONTROL NUMBER	DATE	MODEL	REGISTRATION NUMBER	CESSNA SERIAL NUMBER
237035	02/07/83	PRESS. CENTURION II	N6586W	P21000800

ITEM	WEIGHT	ARM	MOMENT
LIGHTS, STROBE	3.3	38.9	128
RECORDER, FLIGHT HOUR	.5	9.8	5
HARNESS, SHOULDER/SEAT BELT SYS. 3 & 4 (EXCH)	1.2	71.0	85
STABILIZER ABRASION BOOTS	2.7	202.0	545
400 ADF W/BFO R446A (EXCH.)	7.5	27.0	203
3 IN HSI SYS (SLAVED) INSTALLED W/400 NAV IND EXCH	6.6	42.7	282
DME-450C	8.5	88.3	751
DUAL 400 NAVCOMS (RT485B) W/400 VOR/LOC (EXCH)	17.0	13.7	233
SWITCH, REMOTE TRANSPONDER IDENT	.2	17.0	3
OXYGEN SYSTEM, COMPLETE, 6-PORT CANISTER TYPE	5.6	59.1	331
ANTI-ICE SYSTEM, PROPELLER (ELECTRIC)	5.1	22.0-	112-
INCONNEL EXHAUST SYSTEM NEGL.			
BASIC EMPTY WEIGHT SUPERCEDED 4 17 85	[redacted]		
USEFUL LOAD			
MAXIMUM RAMP WEIGHT	4016.0		—
MAXIMUM TAKE-OFF WEIGHT 4000.0			
4-17-85			
Riley International Intercooler	8.0	-3.0	-24
Temp Gage	.25	18.0	4.5
EMPTY WEIGHT	~~2625.75~~	~~40.8~~	~~107142.5~~
USEFUL LOAD	~~1390.25~~		
Revised 9/29/88 EMPTY WT	2619.18	40.46	105977.98
USEFUL LOAD	1396.82		
GROSS WT	4016.0		
Transcal D120-P2 Removed + a Transcal SSD120 installed [redacted] 11-09-07	No change	12.5	No change

Original Weight and Balance with modification (p. 2)

Equipment List

The FAA required Cessna to produce/publish an Equipment List at the time of manufacture. Found in Section 6 of the POH/AFM behind the Weight and Balance data, it lists everything that was or could have been installed on the airplane when it left the factory. The items are coded as follows:

R = Required equipment

S = Standard equipment items

O = Optional equipment items replacing required or standard items

A = Optional equipment items which are in addition to required or standard items

ITEM NO	EQUIPMENT LIST DESCRIPTION	REF DRAWING	WT LBS	ARM INS
	-STROBE LIGHTS IN WING TIP (SET OF 2)	C622006-0107	0.2	38.7
C49-S	LIGHTS, COWL MOUNTED LANDING & TAXI -LIGHT BULBS (SET OF 2)	2113000 GE 4591	2.0* 1.0	-30.0* -34.5
C57-A	FLOOD LIGHT INSTL., VERTICAL TAIL	1231069-1	1.7	226.0
C61-A	LIGHT INSTALLATION, ICE DETECTOR (WING)	2101005-1	0.4	8.2
C73-A	LIGHT, CABIN FLOOD (WHITE)	2101014-1	0.2	70.0
	D. INSTRUMENTS			
D01-R	INDICATOR, AIRSPEED	C661064-0221	0.7	18.0
D01-O	TRUE AIRSPEED INDICATOR	1201108-13	0.8	18.0
D04-R	INSTRUMENT AIR ALTERNATE STATIC SOURCE	2100004	0.3	39.2
D07-R	ALTIMETER, SENSITIVE	C661071-0201	0.9	18.0
D07-O-1	ALTIMETER, SENSITIVE (FEET & MILLIBARS)	C661071-0202	0.9	18.0
D07-O-2	ALTIMETER, SENSITIVE (35,000 FT, 20 FT. MARKINGS & USED IN SOME FOREIGN COUNTRIES)	C661025-0102	0.7	18.0
D07-O-3	ALTIMETER, SENS. (FT & MILLIBARS)	C661071-0204	0.9	18.0
D10-A	ALTIMETER, 2ND INSTRUMENT INSTALLATION	2101013-9	1.1	17.5
D16-A-1	ALTIMETER, ENCODING (REQUIRES RELOCATION OF REGULAR ALTIMETER) -ALTIMETER	2101013-7	3.2* 2.8	15.7* 16.0
D16-A-2	ALTIMETER, ENCODING (REQUIRES RELOCATION OF REGULAR ALTIMETER) (FEET & MILLIBARS) -ALTIMETER	2101013-8	3.2* 2.8	15.7* 16.0
D16-A-3	ALTITUDE ENCODER, BLIND (INSTRUMENT PANEL	2101011-2	1.6*	16.6*

One page from the Equipment List

SECTION 6
WEIGHT & BALANCE/
EQUIPMENT LIST

CESSNA
MODEL P210N

ITEM NO	EQUIPMENT LIST DESCRIPTION	REF DRAWING	WT LBS	ARM INS
E07-O-1	SEAT, CO-PILOT FIXED HEIGHT, ARTICULATING WITH LUMBAR SUPPORT	1214175	19.4	44.0
E07-O-2	SEAT, CO-PILOT VERTICAL ADJUST. AND ARTICULATING	1214175-11	24.3	39.5
E07-O-3	SEAT, CO-PILOT VERTICAL ADJUST. AND ARTICULATING W/LUMBAR SUPPORT	1214175-12	25.4	39.5
E09-S	SEATS, TWO 2ND ROW INDIVIDUAL	1214174	37.5	73.0
E11-S	SEAT, TWO PLACE 3RD ROW BENCH	2114017-2	23.4	104.5
E15-R	SEAT BELT ASSY, PILOT	S-2275-103	1.0	37.0
E15-S	SHOULDER HARNESS ASSY, PILOT	S-2275-201	0.6	37.0
E19-A	INERTIA REEL INSTALLATION, 1ST ROW SEAT BELT & SHOULDER HARNESS (WT NET CHG.)	1201057-2	0.6	89.1
E23-S	BELT & SHOULDER HARNESS ASSY, CO-PILOT	S-2275-3	1.6	37.0
E27-S	BELTS ASSY, 2ND ROW OCCUPANT LAP (SET OF 2)	S-1746-25	2.0	71.0
E27-O	BELTS & SHOULDER HARNESS ASSY., 2ND ROW OCCUPANT LAP (SET OF 2)	S-2275-5	3.2	71.0
E29-S	BELTS ASSY., 3ND ROW OCCUPANTS (SET OF 2)	S-1746-17	2.0	101.0
E29-O	BELTS & HARNESS ASSY., 3RD ROW (SET OF 2)	S-2275-32	3.2	101.0
E34-O	UPHOLSTERY, LEATHER SIDEPANELS NET CHANGE		2.0	65.0
E35-O-1	LEATHER SEAT COVERING, FULL COVER 6 PLACE (ADDS WEIGHT OF 0.5 LB. PER SEAT)		3.0	73.1
E35-O-2	LEATHER SEAT COVERING, PART COVER 6 PLACE		1.5	73.1
E47-A	OXYGEN SYSTEM, SOLID STATE GENERATOR WITH APPROXIMATELY 15 MINUTES USE	2100001-1	5.6*	59.1*

6-22

11 September 1981

Equipment List page showing weight of seats

Why do you care what is on the Equipment List? For starters, if something isn't working in your airplane or something has been removed, you will be able to determine whether or not you can legally fly the airplane without that item either not working or not installed. If it has an "R" at the end of the ITEM NO on the Equipment List, you probably need an FAA Ferry Permit to fly the airplane.

Using this list can assist you in calculating your Weight and Balance before a flight if, for example, you wanted to remove the back seat or a center seat for a trip. The item you want to remove is listed along with the weight in pounds and the moment arm. Having this information readily available makes it very easy to do a calculation.

One of the nice features of the 210 aircraft is that we can remove seats and change our load configuration. Yes, we really can take seats in and out to accommodate our loading desires.

The most common seating configuration I see in 210s is a five-seat configuration, with the center seat behind the pilot removed. This

Five seat configuration — left middle seat removed

allows for much easier boarding for the passengers and gives room to place luggage in the seat position and secure it with the seat belt straps.

The rear bench seat is commonly removed by pilots with dogs and by pilots wanting to carry ice chests full of fish back from a trip. A caution here: some 210s have avionics under their back seats! Have a look when you remove the seat. You may have to make a removable cover to protect the avionics with the seat removed.

I have had my FBO make me four different Weight and Balances for various seating configurations that I use:

- All six seats installed
- Front and center seats installed, rear bench seat removed
- Front seats installed, one center seat removed, rear bench seat installed
- Front seats installed, one center seat installed, rear bench removed (this configuration is my favorite for fishing trips)

Having these four Weight and Balances already done by the FBO and in plastic sleeves in Section 6 of the POH makes it very easy for me to calculate my actual Weight and Balance using the worksheets found on page 6-10 of the POH, regardless of the seat configuration I have in the airplane for the intended trip.

Caution: The use of electronic/spreadsheet Weight and Balance computers/programs and "apps" can land you in trouble without you realizing it. If you are going to use a spreadsheet program or an app that someone has written for their airplane, you need to validate that program for your own airplane. Don't trust that what you get from someone else that might work for their airplane is going to work for you in your airplane. By validating, I mean you need to run a least a half-dozen different sample Weight and Balance calculations using the program and then manually run the same samples using the charts in Section 6 of your POH/AFM. Verify.

I see pilots doing a takeoff Weight and Balance, but I rarely see a pilot perform an anticipated landing Weight and Balance. When you

SECTION 6
WEIGHT & BALANCE/
EQUIPMENT LIST

CESSNA
MODEL P210N

SAMPLE LOADING PROBLEM	SAMPLE AIRPLANE		YOUR AIRPLANE	
	Weight (lbs.)	Moment (lb.-ins. /1000)	Weight (lbs.)	Moment (lb.-ins. /1000)
1. Basic Empty Weight (Use the data pertaining to your airplane as it is presently equipped. Includes unusable fuel and full oil)	2496	101.8		
2. Usable Fuel (At 6 Lbs./Gal.) Standard Tanks (87 Gal. Maximum) . . .				
Reduced Fuel (64 Gal.)	384	16.5		
3. Pilot and Front Passenger (Station 34 to 46) .	340	12.6		
4. Center Passengers (Station 59 to 76)	340	24.1		
5. Aft Passengers	340	34.3		
6. *Baggage (Area "A") (Station 124 to 152) 200 Lbs. Max. . .	92	12.7		
7. *Baggage - Aft (Area "B") (Station 152 to 166) 80 Lbs. Max. . . .	24	3.8		
8. RAMP WEIGHT AND MOMENT	4016	205.8		
9. Fuel allowance for engine start, taxi and runup	-16	-.7		
10. TAKEOFF WEIGHT AND MOMENT (Subtract step 9 from step 8)	4000	205.1		
11. Locate this point (4000 at 205.1) on the Center of Gravity Moment Envelope. <u>Since this loading falls within the shaded area of the moment envelope, proceed with steps 12, 13 and 14.</u> If the computed loading point falls within the clear area of the moment envelope, no further steps are required and the loading is assumed satisfactory for take-off and landing provided that flight time is allowed for fuel burn-off to a maximum of 3800 pounds before landing.				
12. Estimated Fuel Burn-Off (Climb and Cruise) (43 Gallons at 6 Lbs./Gal.)	-258	-11.1		
13. Subtract step 12 from step 10 for estimated airplane landing weight	3742	194.0		
14. Locate this point (3742 at 194.0) on the Center of Gravity Moment Envelope. Since this point falls within the overall envelope, the loading may be assumed acceptable for landing. *The maximum allowable combined weight capacity for baggage areas A and B is 200 lbs.				

Figure 6-5. Sample Loading Problem (Sheet 1 of 2)

6-10

Original Issue

CESSNA
MODEL P210N

SECTION 6
WEIGHT & BALANCE/
EQUIPMENT LIST

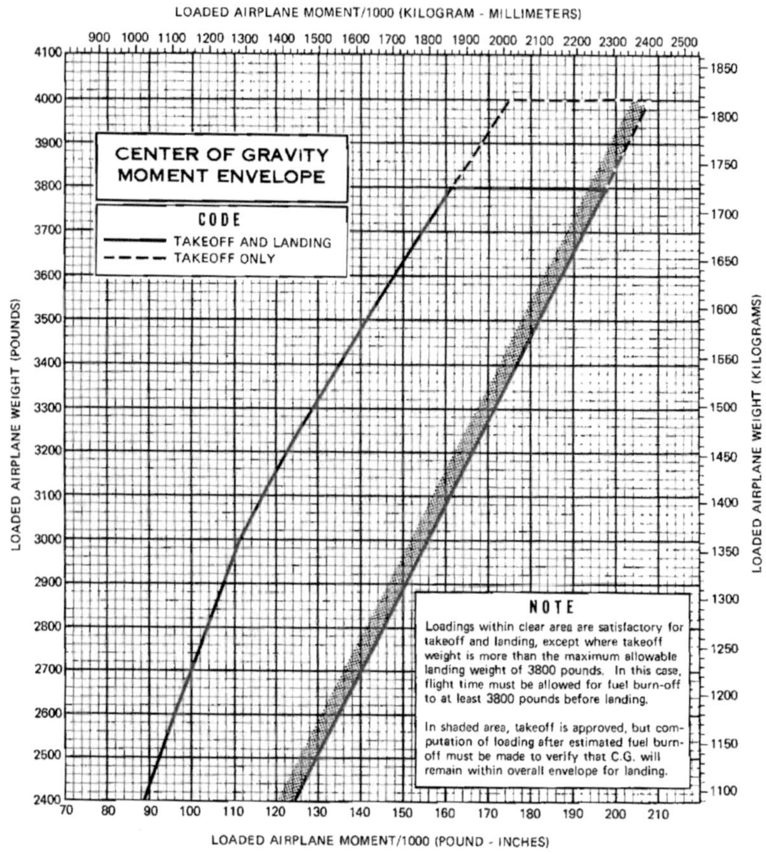

Figure 6-7. Center of Gravity Moment Envelope

get your 210, please perform both a takeoff and landing Weight and Balance enough times so you will know with certainty that you are not going to end up with an excessively aft CG for landing. It is hard to be able to take off with the CG within range in a 210 and not be able to stay within the CG range on landing, but it can be done.

From personal experience, I know that failing to correctly perform a landing Weight and Balance for a flight can result in a CG so far

SECTION 6
WEIGHT & BALANCE/
EQUIPMENT LIST

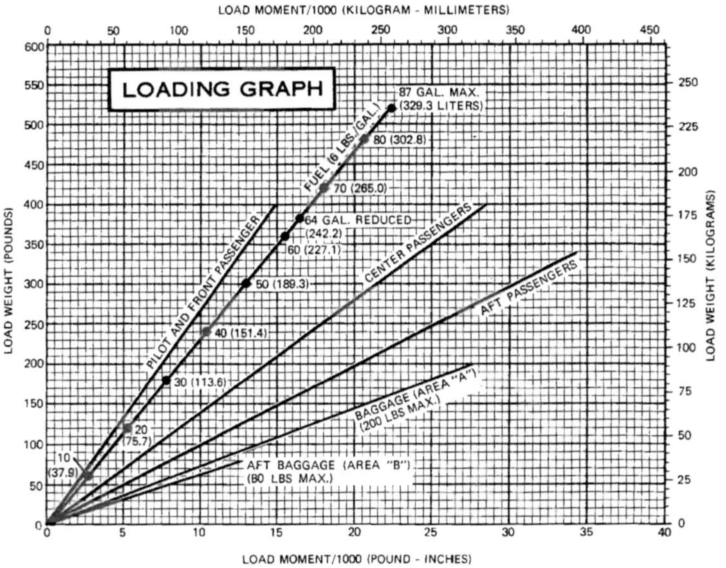

NOTE: Lines representing adjustable seats show the pilot or passenger center of gravity on adjustable seats positioned for an average occupant. Refer to the Loading Arrangements diagram for forward and aft limits of occupant C.G. range.

Figure 6-6. Loading Graph

CESSNA
MODEL P210N

aft that you can be unable to keep the nosewheel on the ground after landing, even with full forward elevator. That's not a good thing, especially since the nose strut is so extended that the nose gear steering links won't engage.

In Greenland — Refueling in the rain from 55-gallon drums! Wait an hour and then drain the sumps.

5

Preflight Inspection

Your safety depends on it!

Do you know a pilot that pulls out his POH and walks around his airplane reading out of the book as he performs the preflight? Well, there are probably a few pilots that do that, but I haven't met them yet. This is a good thing in one respect, as the POH serves as a checklist and reading it shows the pilot is trying to be thorough. But it is not so good in the long run, as the POH preflight inspection checklist probably doesn't cover all the items you need to check on your airplane before flight. What? How can that be? That *probably* is because the airplane has had some equipment or avionics removed and others put in their place, or new equipment or systems installed.

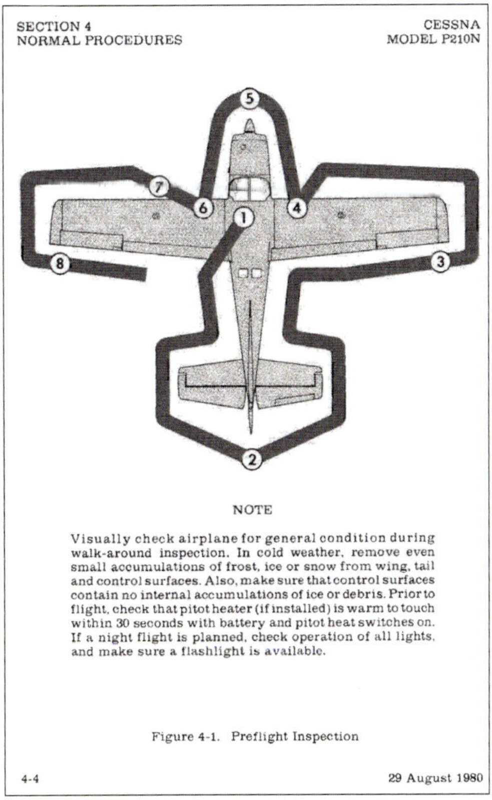
SECTION 4
NORMAL PROCEDURES

CESSNA
MODEL P210N

NOTE

Visually check airplane for general condition during walk-around inspection. In cold weather, remove even small accumulations of frost, ice or snow from wing, tail and control surfaces. Also, make sure that control surfaces contain no internal accumulations of ice or debris. Prior to flight, check that pitot heater (if installed) is warm to touch within 30 seconds with battery and pitot heat switches on. If a night flight is planned, check operation of all lights, and make sure a flashlight is available.

Figure 4-1. Preflight Inspection

4-4

29 August 1980

POH preflight inspection diagram

Preflight Inspection is found in the Normal Procedures (Section 4) of the POH or AFM, and in Section 1 of the Owner's Manual. This is the place to start, but it is only the beginning.

Simple things like installing inflatable door seals add an item to the preflight. Pulse lights or wingtip lights such as those RMD makes for the 210 should be checked during the preflight. Since Cessna never offered them as options, you won't find information regarding them anywhere in the POH. You have to look for the manufacturer's POH/AFM supplement or operating instructions.

How about sitting down with the POH at your computer and making a list of all the equipment that is actually in your airplane today? Go to Section 4 of the POH and use it as a guide to create your own preflight checklist. After you have done that, go to Section 9 of the POH, read the flight manual supplements for all the installed equipment that needs to be preflighted, and add those items to your draft preflight checklist.

T210 with Avidyne Glass Panel installed—the only one of these made!

Use the POH preflight inspection as a guide to add and delete the items that apply to your airplane. The more "good stuff" you have in your airplane, the more detailed the preflight needs to be and the longer this checklist will be.

If you have an IFR GPS installed, you have to check that the database is current. If you have a state-of-the-art autopilot installed, you will need to read the autopilot's preflight inspection portion of the Flight Manual Supplement to complete your checklist. As an example, some new autopilot installations add additional static ports in the fuselage of the 210 and, obviously, they need to be checked as well as the factory-installed pitot system static ports.

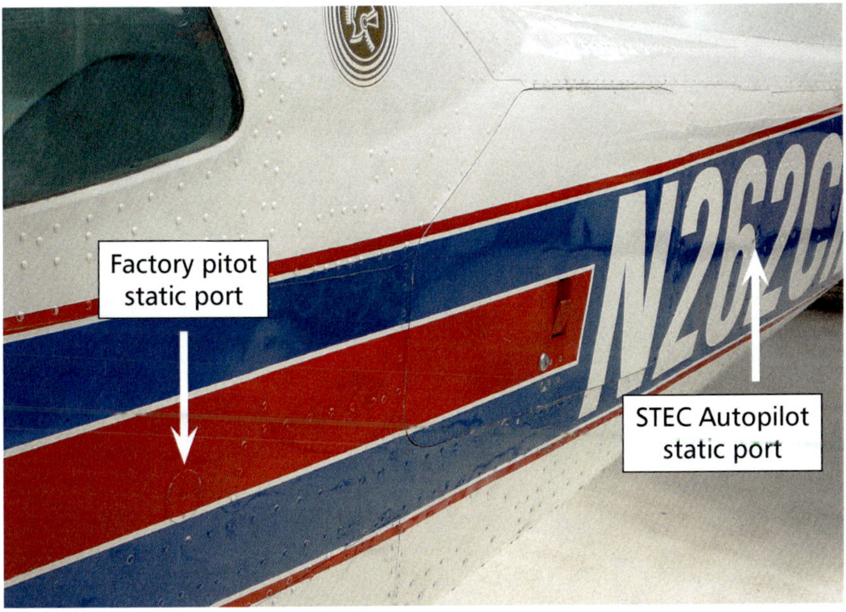

Aircraft pitot static port and additional autopilot static port

If you have speed brakes or spoilers installed, you need to activate them before you walk around the airplane so you can see that they both are up when they are supposed to be; and at some point you have to put them back down, then look on top of the wings and make sure they are in fact properly stowed. You also need to verify that the warning light in the cockpit works. If you have speed brakes installed, you have to inspect the speed brake drain tubes on the underside of the wings for security and proper length. (Never push up on the bottom of the tube as you can eventually dislodge them and thereby defeat their purpose of draining all the water from their holding container.)

For those aircraft equipped with VGs, it is required that you check to made sure that some of the little vortex generator "fins" have not fallen off the aircraft before flight. The POH Supplement tells you exactly how many may be missing from each surface area before you are prohibited from flying the aircraft.

Vortex generators

As you move through the creation of your checklist, put things in the order of flow across the cockpit and around the airplane so that you aren't reaching back and forth across the cockpit as you check and set switches and dials, lower the flaps, and so on, or stopping and backing up as you move around the exterior of the aircraft.

Once you think your checklist is pretty complete, print out a copy and go perform the preflight on your airplane using your newly created checklist. You will probably find that you need to make a few modifications, such as moving a few things around and also adding

CENTURION N54XXX P-210/G — **PAGE 1**

PREFLIGHT CHECK

1. Fuel Tanks ... COLLARS OR FULL?
2. Oil Level ... CHECK ... ADD if less THAN 8 QTS
3. Tire Pressures ... Nose 88 Mains 55

Inside the Cockpit

1. Gear Lever ... DOWN
2. Ignition Switch ... OFF
3. POH ... in the center console?
4. Hydraulic Fluid Level ... CHECK
5. Control Lock ... REMOVE
6. Throttle Lock ... REMOVE
7. Cowl Flaps ... OPEN
8. Cabin Heat ... OFF
9. Air Conditioner ... OFF
10. Standby Generator ... OFF
11. Avionics ... OFF
12. Standby Vacuum Pump ... OFF
13. Alt Static Source ... IN
14. Aux Fuel Pump ... OFF
15. Circuit Breakers ... CHECK IN
16. **Master Switch** ... ON
17. Avionics Fan ... CHECK
18. Suction Gauge Buttons ... VISIBLE
19. Test Standby Vacuum Pump, then turn ... OFF
20. Fuel Gauges ... CHECK QUANTITY
21. Fuel Selector ... LEFT or ... FULLER TANK
22. Fuel ... SET Fuel Computer
23. Tach ... RECORD

24. Cabin Pressurization:
a. Dump Valve Control ... IN ... (OUT warm days)
b. Cabin Pressure Switch ... ON ... (OFF warm days)
c. Cabin Alt Selector ... SET ... (High warm days)
d. Cabin Alt Warning Light ... TEST

25. Gear Warning Light ... CHECK
26. Gear Warning Horn ... CHECK
27. Flaps ... DOWN
28. Lights ... ALL ON

QUICK WALK AROUND AIRCRAFT

29. Lights, Strobes, Stall Warning, Pitot Heat ... CHECK

29. **Master Switch** ... OFF

EXTERIOR INSPECTION

Left Fuselage

1. Left Gear Leg, Brake ... CHECK
2. A/C drain tube ... CLEAR
3. Pilot Static Source ... CLEAR
4. Belly antennas ... SECURE
5. A/C belly vent ... CLEAR
6. A/C fuselage vents ... CLEAR

Empennage

1. Boots ... Good no damage
2. Grounding Strap ... SECURE
3. Static Wicks ... All SECURE
4. Tail Tie Down ... REMOVED
5. Control Surfaces ... Proper Movement
6. Elevator/Rudder ... SECURE
7. Trim Tab Castle Nut ... SECURE
8. Static Wicks ... SECURE
9. Grounding Strap ... SECURE

Right Fuselage

1. Pilot Static Source ... CLEAR
2. Upper and Lower Antennas?
3. Windows, Emergency Exit?

Right Wing

1. Lateral Flap Travel ... CHECKED
2. Wing Fuel tank ... SUMP
3. Gear leg, brakes ... CHECK
4. Side windows ... CHECK
5. Emergency Door ... CHECK
6. Aileron Gap Seal ... CHECK
7. Aileron ... CHECK
8. Static Wicks ... SECURE
9. Fuel Tank Vent ... CHECK
10. Flint Tip Tank ... SUMP
11. Boots ... CHECK
12. Tie down ... REMOVE
13. Gear Mirror ... Check SECURE
14. Fuel Tank ... Verify Quantity
15. Fuel filler cap ... SECURE
16. Belly fuel drain ... SUMP
17. Windshield ... CHECK
18. Intercooler inlet ... CLEAR
19. Cowl door ... SECURE
20. Exhaust pipe ... SECURE

PAGE 2

Nose

1. A/C belt ... SECURE
2. Propeller ... Check or Nicks
3. Anti-ice Boots ... SECURE
4. Spinner ... no cracks, SECURE
5. Landing/Taxi Lights ... SECURE
6. Oleo Strut Extension ... Proper
7. Nose gear tunnel ... CLEAR
8. Nose gear doors ... SECURE
9. Oil Cooler Opening ... CLEAR
10. Left Cowl Door ... SECURE
11. Windshield ... CHECK
12. Hot Plate ... SECURE
13. Heat Exchanger Duct ... CLEAR
14. Engine Fuel Sump ... SUMP

Left Wing

1. Belly fuel drain ... SUMP
2. Boots ... CHECK
3. Fuel Tank ... Verify Quantity
4. Fuel filler cap ... SECURE
5. Flint Tip Tank ... SUMP
6. Fuel Tank Vent ... CHECK
7. Static Wicks ... SECURE
8. Aileron Gap Seal ... CHECK
9. Aileron ... CHECK
10. Wing Fuel Tank ... SUMP
11. Gear Leg, brakes ... CHECK

Baggage Door

77. Pre-flight Materials ... SECURE
78. Baggage Door ... SECURE

Preflight Checklist N454XXX new.doc

Example preflight checklist

a thing or two. Now, go back home and make the necessary changes that you feel are needed.

Once you are fairly certain that you have a good checklist, print out two fresh copies, take them to the local copy store, and have them laminated. Ask for 10-mil lamination and you will get a nice stiff checklist that doesn't bend easily and will be very durable. One copy is to use for your regular preflight; the other is to have holes punched in it and stored in the front of the POH — for when you lose the one you usually use.

Some owners like to put Velcro on the preflight checklist that they regularly use and attach it in the baggage compartment when not in use. The baggage compartment is where they normally start their preflight because that is where their rubber gloves and fuel tester are stored in a little snap-top plastic container, along with a cloth or foam pad to kneel down on during preflight. You can't properly preflight a Cessna 210 without getting some dirt on your hands and knees.

Let's talk about that small plastic container in the baggage compartment before we go on. In it we find things like windshield cleaner, a mild degreaser such as Fantastic or some other like product that we use to clean off oil, bugs, etc. We also find in it a quart or two of oil, some disposable paper oil funnels and paper towels, a fuel measuring tube, and a fuel tester. That fuel tester should have a metal shank so that when it is pushed up inside the fuel drains it doesn't leave little plastic shavings in the drain valve. We also need clean towels in the container in their own Ziploc one-gallon baggies so they aren't contaminated by the various chemicals in the box before we use them on

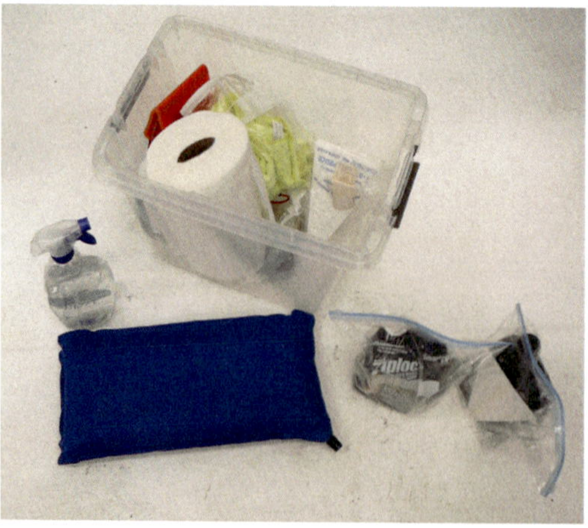

Preflight box

the windshield. Never use paper products on the windshield, as they will make little scratches in the Plexiglas.

One of the most commonly missed items I see is not checking the aircraft's fuel quantity gauges before going out and verifying the quantity by visually checking the tanks. Many pilots use an aftermarket clear plastic tube with gallon markings on it to determine exactly how much fuel they have. Those tubes are available at most pilot shops for less than $20. You will have to calibrate the tube, as I have yet to see one for the various 210 fuel tank sizes.

Measuring the fuel quantity

The fuel system in a Cessna 210 has been blamed for a number of accidents. Many of those accidents have involved an engine failure due to fuel exhaustion. A review of NTSB Accident Reports shows that on average almost one Centurion accident a month occurs from fuel mismanagement or fuel exhaustion!

NTSB Identification: CEN10LA512
14 CFR Part 91: General Aviation
Accident occurred Monday, August 30, 2010 in Houma, LA
Probable Cause Approval Date: 05/11/2011
Aircraft: CESSNA 210, registration: N210PR
Injuries: 1 Minor, 1 Uninjured

> During the first leg of a cross-country flight, the pilot refueled the airplane with about 55 gallons. The flight then proceeded to the destination airport and experienced a total loss of power about 3 hours later. The pilot elected to conduct a forced landing to a cane field, during which the airplane impacted the ground hard and nosed over, coming to rest inverted. The pilot reported that the airplane normally burned about 17 gallons an hour and that he expected to land at the destination airport with about 32 gallons remaining. A visual inspection of the airplane and accident site did not identify any signs of a fuel spill. A post-accident examination of the airplane's fuel tanks and system revealed only trace amounts of fuel. Additionally, the examination of the engine and airframe revealed no apparent leaks or abnormalities that would have prevented normal operation.
>
> The National Transportation Safety Board determines the probable cause(s) of this accident as follows:
>
> The pilot's improper fuel management, which resulted in a loss of engine power due to fuel exhaustion.

The POH cautions the pilot to ensure the aircraft is level when checking fuel quantities. In addition to the wings being level, the pitch attitude of the airplane is also important so that you do not inadvertently leave air pockets in the fuel tanks. I gently rock the wings of my Centurion when refueling to help get all the air out of the tanks.

There are a lot of fuel drains in your Centurion. In addition to the wing fuel drains and the sump drain, reservoir tanks are located under the fuselage on both sides of the airplane in the pre-'82 Centurions; these should be drained as part of the preflight. There is only one belly drain on the '82 and later Centurions. If you have Flint tip tanks and/or a baggage fuel tank, those additional fuel drains need to be incorporated into your preflight check as well. That could be four more fuel drains to check.

Fuel drains under the fuselage

As you perform your prefight of your turbocharged or pressurized Centurion, don't be surprised to see a small puddle of what looks like thinned out oil below the right cowl flap.

Residual oil pool under a turbocharged 210

All the turbocharged engines expel residual oil from the oil breather tube after a flight. You will notice that after a long, high-altitude flight, the puddle will be larger than from a shorter or lower-altitude flight. Get used to the size of the oil puddle so that if it is abnormally large you can ask your mechanic to check for a problem.

Cessna does not instruct you to lower your flaps during preflight in most POHs, but you must extend them at least to 20° to check for excessive lateral flap movement. To conserve the battery, I don't like to run the flaps back up until after I have started the engine, but they are a head-knocker when deployed if you aren't careful. Wearing a hat during preflight has saved me a lot of scalp damage.

If your aircraft has gear doors, you are going to have to manually pump them down to properly complete your preflight inspection. The POH explains the process. And it is one that involves pulling and then resetting a circuit breaker, so it needs to be on a checklist. Taxiing out for takeoff with your gear doors hanging open has spawned a variety of comments from other pilots on the radio.

Gear doors closed

Gear doors open during preflight

There are a number of antennas on these airplanes, with many of them underneath the aircraft. While you are under the airplane draining the fuel sumps/tanks, check the antennas as well. The antennas on the turbocharged airplanes are going to get some oil on them which reduces their efficiency. If they have oil on them during your preflight, wipe them clean with that cleaning solution and the paper towels you have in your little plastic box. You are going to get a little dirty preflighting a Centurion, so just plan on it.

Numerous antennas under the fuselage

If you have air conditioning, depending on the system installed, you may have a drain under the aft fuselage to check, as well as vents on the side of the fuselage and under the belly. If these aren't on your preflight checklist, you will probably miss them, especially if you are interrupted during the preflight. This seems to happen with great regularity during my preflight.

NTSB Identification: CEN09CA125

14 CFR Part 91: General Aviation

Accident occurred Saturday, January 10, 2009 in Albuquerque, NM

Probable Cause Approval Date: 05/06/2009

Aircraft: CESSNA T210L, registration: N211LC

Injuries: 1 Uninjured

Shortly after takeoff and following the landing gear retraction, the pilot realized that he failed to remove the tow bar from the airplane's nose landing gear. After multiple failed attempts to extend the airplane's nose landing gear, the pilot elected to retract the main landing gear and perform a gear-up landing. The airplane's fuselage and right aileron sustained damage during the landing.

> The National Transportation Safety Board determines the probable cause(s) of this accident as follows:
>
> The pilot's failure to remove the airplane's tow bar before takeoff.

Checking alternator belts usually requires removing a top cowling, and few folks are going to be bothered with that. If you are about to depart on a particularly high-risk flight, such as hard IFR or a long leg over water, why don't you take the time to remove that left cowling and check inside thoroughly? Check not just the alternator belt and ground wire, but the fuel spider as well. Why take a chance when it takes only five more minutes to be safer?

TSIO-520R engine

While you have the cowling off, if you have an electric standby vacuum pump installed, check the security of the pump (it will be mounted on top of the battery box) and the related hoses.

Have you noticed that the 210 main gear tires are wearing on the outside edges? That is partially because of the way the gear is angled

Electric standby vacuum system mounted on top of the battery box

before you put weight on it while landing, but it can also be due to improper tire pressure, especially low pressure. Checking tire pressures is one of the most often-overlooked items I see on a Centurion preflight. Your tire pressure gauge should be in that plastic preflight equipment box just like the fuel tester, so check those tires during every preflight. Some pilots have a little electric air compressor to pump up the tires in the baggage compartment that plugs into the cigar lighter; other pilots have an air tank or compressor in their hangar. (Before you plug your little auto air compressor into your aircraft cigar lighter outlet, check the voltage as it may have been converted to 24 volts; if it has been, it will burn out the 12-volt automotive air compressor very quickly.)

If you have "cupping" on your nose tire — little symmetrical indentions — that is caused by failing to pull your yoke back to take the weight off the nosewheel when taxiing. If you don't pull it back, you are literally grinding the rubber off the side of the tire as you make the turn. This is found most often on P210s, but I have seen it on all three Centurion models.

If you are going to be flying over or near water, ensure your life preservers are within their expiration date and that everyone is either wearing them or knows how to put them on and is physically able to do so after they are in their seats—before you take off.

If this flight is to be over water or into/over rugged terrain, have you briefed and shown them how to use their life jackets, or the location of and how to access and use the survival equipment?

For flights that might require oxygen use, check each oxygen cannula or mask while on the ground to ensure it is working properly, and then make sure your passengers know how to use it during flight.

Speaking of your passengers, have you briefed them on the essential information they need for the flight? Have you explained the hand signal you use when you want them to stop talking so you can listen to the aircraft radio and attend to your pilot duties? Have you told them, "No talking during takeoff and landing"? Have you taken out your Cessna-prepared passenger briefing card, had them read it, and then asked if they have questions?

If you don't have passenger briefing cards like the examples shown, they are available from the Cessna Customer Service Center.

Your survival equipment should be up to date, suitable for the route and time of year of the flight, and located so that it can be immediately retrieved should it be needed. Don't store your survival kit under the luggage in the baggage compartment; put it right in front of the baggage door opening if you aren't carrying it in the cockpit.

As you can see, a thorough preflight of a well-equipped Centurion can take a little time. Don't forget to arrive at the airport with plenty of time to preflight before your departure so you aren't rushing.

Take the time and effort to perform a thorough preflight every time. Your life and the lives of your passengers are worth the time it takes to do it right every time. Make it a habit.

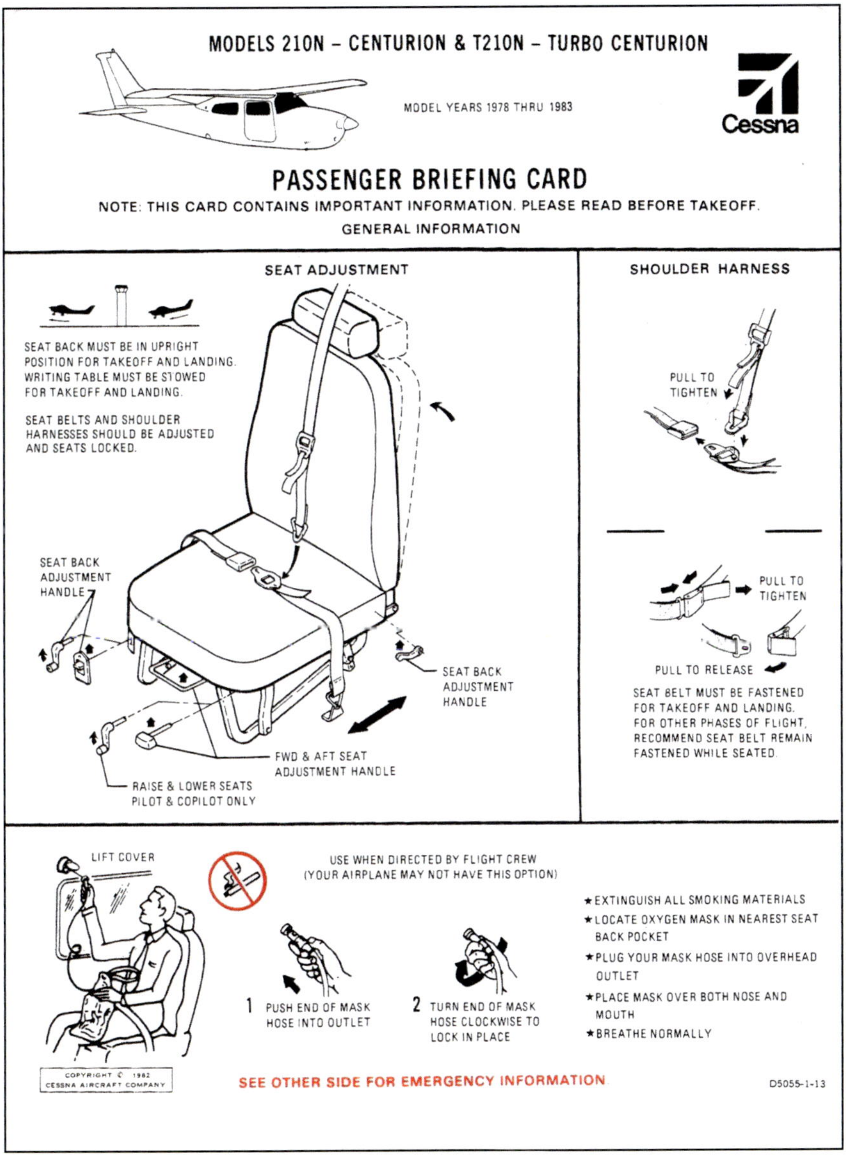

Passenger briefing card side 1

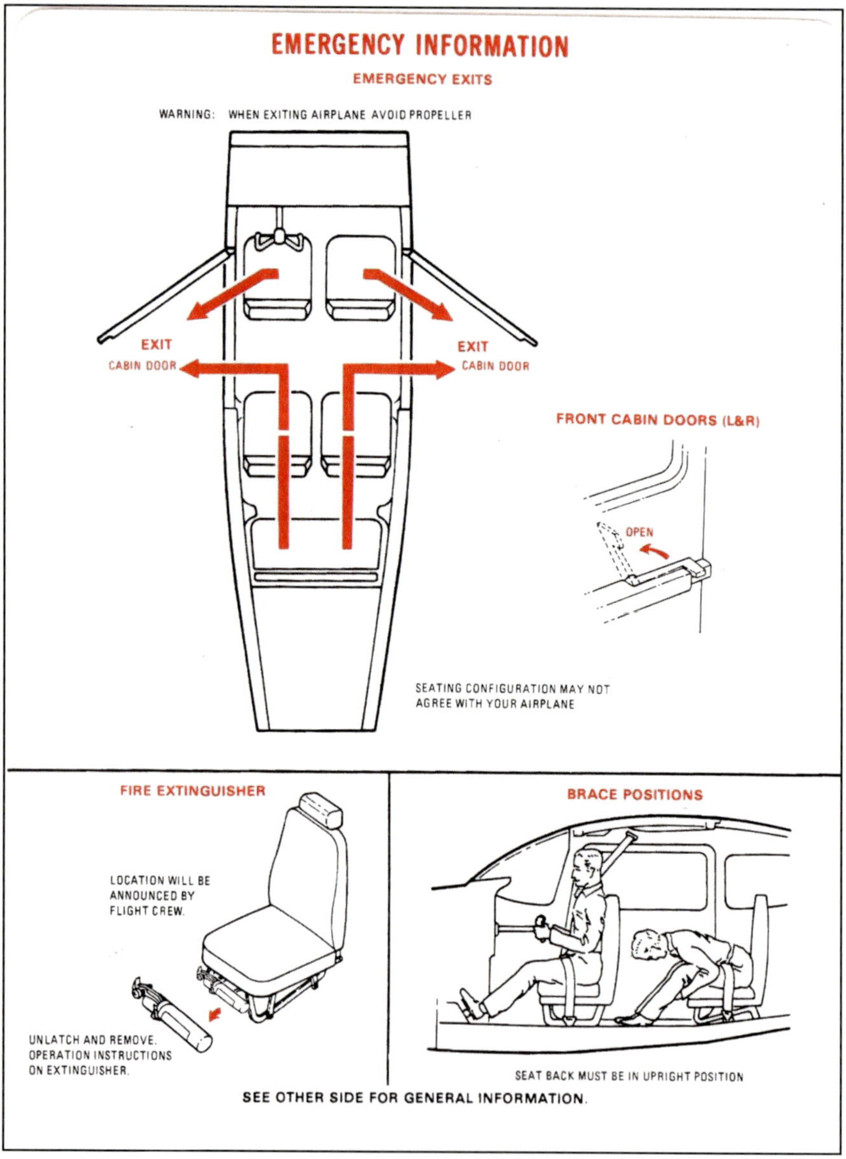

Passenger briefing card side 2

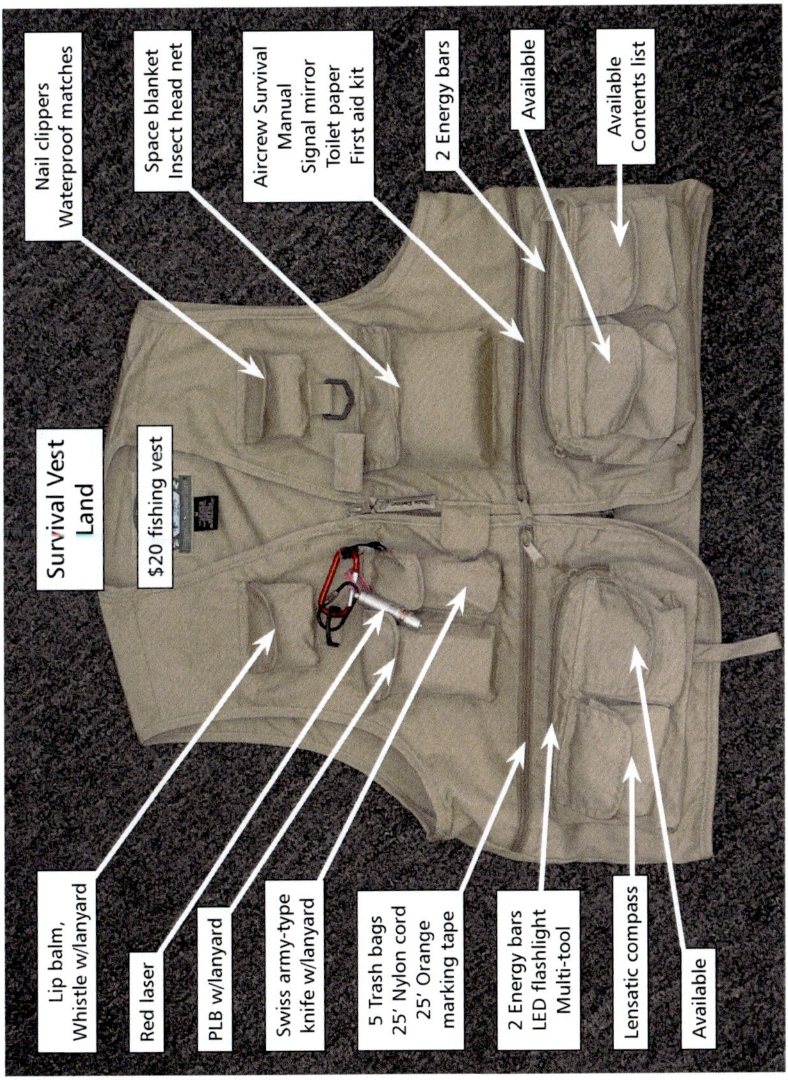

An inexpensive land survival vest

Intercooler-equipped P210N at Burns, Oregon airport

6

Preparation for Flight

Planning, Performance, Taxi, and Run-Up

According to many aviation journalists who have flown the aircraft, the Centurion is the highest pilot-workload aircraft of any single-engine aircraft manufactured. Yet most of them also say it is a nice-flying airplane. Are they being inconsistent with these statements?

First they say it is stable, and then they say it is like flying a truck. Well, those statements probably aren't inconsistent. We refer to our T210 as a flying Mercedes station wagon, not a truck. The lighter the airplane, the less truck-like it is. A lightly loaded 210 gets off the ground and climbs quickly. The P210, especially near gross weight, never leaps off the ground, nor does it climb quickly. The P210 is certainly more solid, or truck-like, than the Cessna 210 or T210, as it is heavier and the airframe is stiffer.

The Centurion is a high-altitude, cross-country IFR aircraft. We want it to be stable. We want this 3,800- to 4,100-pound airplane to carry six people and their baggage at altitudes up to 25,000 feet or even higher. When flying it IFR, we don't want it bouncing around like a cork on the ocean, and it doesn't; the Centurion is a very stable aircraft in flight.

What makes for high pilot workload? A lot of systems raise the workload. The speed of the aircraft increases workload because it gives us less time to accomplish required tasks. The type of flying we are doing changes our workload. A Sunday afternoon VFR flight is really easy compared to a long IFR flight with icing — it doesn't matter what airplane you are flying.

The Cessna 210 is overall fairly simple, but it can be made very complex too. It has a big engine, retractable gear, a constant-speed pitch prop, and flaps. It is your standard complex single-engine airplane. But a number of systems can be added to make it quite complex. The T210 adds complexities with its turbocharged engine and built-in oxygen. The P210 adds pressurization, usually air conditioning, and in many cases known ice equipment.

As the number of systems increase, the pilot's workload increases. Any of these models can be certified for known ice, with either boots or TKS systems, so there are more complexities — not just in systems but in the operation of those systems that are used in already challenging flight regimes.

Additional fuel tanks, such as Flint Tip Tanks or a baggage compartment supplemental fuel tank, add complexity as they require the pilot to check more gauges and operate the transfer pumps within certain parameters of flight. This adds to the pilot workload. We can add backup alternators, GPS systems, flight director and autopilot systems, and glass panels. The list goes on and on.

Overall, the workload of the basic aircraft isn't much more than any other high-performance, retractable gear, single-engine airplane. It's just that we add all kinds of systems and their related complexities to the airplane and then say the airplane has a high workload. Well, any single-pilot aircraft with a turbocharged engine, pressurization, air conditioning, de-ice equipment, etc. *is* going to have a higher workload than one without all those items.

One of the reasons we want this big six-place airplane is, of course, to carry passengers. But passengers increase workload because they have to be briefed on their conduct in the airplane, use of oxygen if the aircraft is so equipped, egress, etc. Passengers always have questions and comments during the flight, and it seems like they usually interrupt our thoughts when we need to be totally concentrating on flying.

If your aircraft has a lot of add-on systems which have been added over time, they may not be laid out on the panel to be pilot friendly. You may have an indicator on one side of the panel and the switches that operate it all the way across the cockpit. That increases pilot workload.

A confusing panel

If you just did a major avionics upgrade and created a custom panel that helped reduce the size of your scan by placing all the indicators, switches and circuit breakers in easy-to-see and reach locations, your workload probably went down!

A clean, well-laid-out panel

So what is the workload of actually flying the airplane? It is not unlike a heavy Cessna 182 with wheels that go up and down. The controls are heavy and stiff, but the aircraft is very stable and totally predictable when we apply control inputs, adjust power, trim, or change configurations.

Planning the Flight

As with any other airplane, our flight will have three phases: departure, en route, and arrival.

Depending upon the flight, your planning may be very simple, such as staying at your home airport and working on takeoff and landing techniques. This we should do regularly, because as with all cross-country airplanes, pilots have a tendency to use them only on trips; after all, that is what the airplane was designed to do! However, failure to regularly practice the basics can leave us wanting when we encounter various weather and runway conditions on our adventures. My school of thought is that you should not perform touch and goes in this or any other retractable gear airplane if your intent is to practice takeoff and landing techniques. Touch and goes do not replicate taking off and landing in high-performance airplanes. There are too many things to do and things are happening too fast for you to be able to replicate an actual takeoff or an actual landing and rollout. (See the article "Touch and Go's" on my website for further information at www.SafeFlightIntl.com.)

Most of our flights in a Centurion are not as simple as staying at the home airport. We are off on a business trip or a vacation with the family under varying weather conditions, into simple or complex airport environments, and frequently under FAA radar control. Therefore, thorough pre-flight planning is necessary for any trip out of the local area.

Weather planning needs close attention. Although these are very capable IFR aircraft, they were not made to penetrate thunderstorm activity.

A few essential items for flight planning

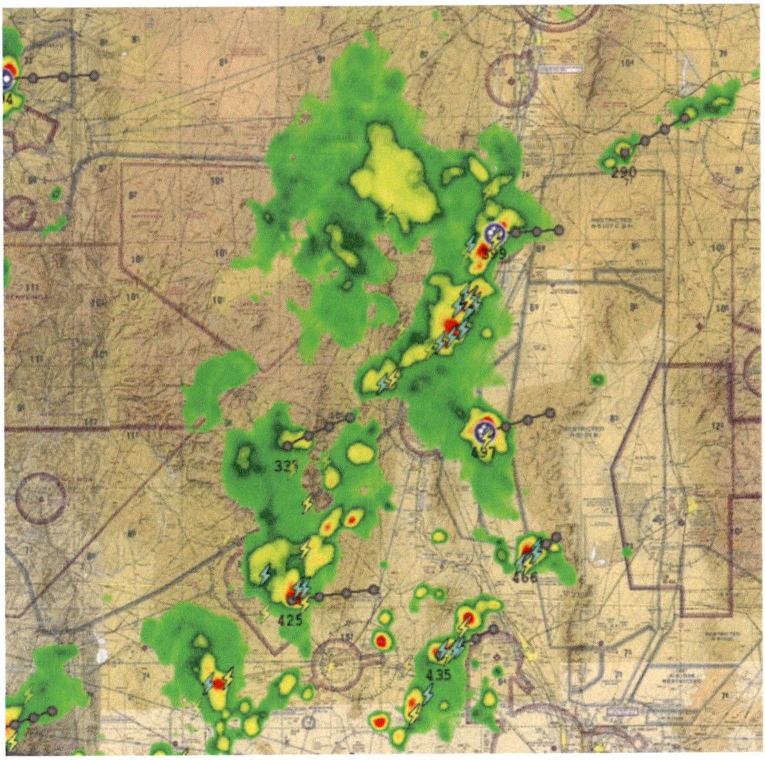

Maybe we should fly around this weather!

In Chapter 1, I mentioned in-flight airframe failures. When you read about one accident below and then review the selected accident reports in the Appendix section, you will see that forecast and actual moderate to severe turbulence and loss of control by pilots (sometimes following loss of flight instruments) in IMC (instrument meteorological conditions) account for the majority of the airframe failures. Sometimes the pilot manages to land the aircraft, others times not.

NTSB Identification: CH106MA115

Accident occurred Wednesday, April 19, 2006 in Ludville, GA
Probable Cause Approval Date: 09/27/2007
Aircraft: Cessna 210A, registration: N6579X
Injuries: 1 Fatal

The airplane flew into an area of severe thunderstorms identified as a mesoscale convective system (or "MCS") with intense to extreme intensities during cruise flight at 11,000 feet then descended rapidly and impacted the terrain. The on-scene investigation revealed no preimpact mechanical malfunctions or anomalies that would have prevented the normal operation of the airplane or its systems.

The airplane entered the severe convective weather; the pilot then requested and received clearance from the air traffic controller to initiate a turn to escape the weather. The airplane was lost from radar about 30 seconds after the pilot initiated the turn. Before the airplane entered the weather, the controller's radar scope depicted a band of moderate to extreme weather along the accident airplane's projected flight path that was consistent with an embedded, heavy-precipitation, supercell-type thunderstorm; however, the I controller did not provide the pilot with any severe weather advisories and did not advise the pilot of the weather depicted on his radar scope.

The National Transportation Safety Board determines the probable cause(s) of this accident as follows:

The pilot's failure to obtain updated en route weather information, which resulted in his continued instrument flight into a widespread area of severe convective activity, and the air traffic controller's failure to provide adverse weather avoidance assistance, as required by Federal Aviation Administration directives, both of which led to the airplane's encounter with a severe thunderstorm and subsequent loss of control.

NTSB Identification: NYC68F0397

14 CFR Part 91 General Aviation

Aircraft: Cessna 210, registration: N6171R

FILE	DATE	LOCATION	AIRCRAFT DATA	INJURIES F S M/N	FLIGHT PURPOSE	PILOT DATA
3-0724	68/3/13 TIME – 1055	WILKES-BARRE, PA	CESSNA 210 N6171R DAMAGE-SUBSTANTIAL	CR– 0 0 1 PX– 0 0 0 OT– 0 0 0	NONCOMMERCIAL BUSINESS	**PRIVATE, AGE 20, 2783 TOTAL HOURS, 257 IN TYPE, NOT INSTRUMENT RATED**

TYPE OF ACCIDENT
TURBULENCE
AIRFRAME FAILURE: IN FLIGHT

PHASE OF OPERATION
IN FLIGHT: NORMAL CRUISE
IN FLIGHT: NORMAL CRUISE

PROBABLE CAUSE(S)
PILOT IN COMMAND – **CONTINUED FLIGHT INTO KNOWN AREAS OF SEVERE TURBULENCE**
WEATHER – TURBUELENCE, ASSOCIATED W/CLOUDS AND/OR THUNDERSTORMS
AIRFRAME – WINGS: SPARS
MISCELLANEOUS ACTS, CONDITIONS – BUCKLED
MISCELLANEOUS ACTS, CONDITIONS – OVERLOAD FAILURE
WEATHER BRIEFING – BRIEFED BY FLIGHT SERVICE PERSONNEL, BY PHONE
WEATHER FORECAST – FORECAST SUBSTANTIALLY CORRECT

SKY CONDITION
UNKNOWN/NOT REPORTED

CEILING AT ACCIDENT SITE
1400

VISIBILITY AT ACCIDENT SITE
5 OR OVER(UNLIMITED)

PRECIPITATION AT ACCIDENT SITE
SNOW SHOWERS

OBSTRUCTIONS TO VISION AT ACCIDENT SITE
NONE

TEMPERATURE–F
26

TYPE OF WEATHER CONDITIONS
VFR

TYPE OF FLIGHT PLAN
NONE

REMARKS – **PLT FLEW INTO KNOWN MDT TO SEVERE TURB.BOTH WINGS DAMAGED.LANDED SAFELY.USWB SIGMET BRAVO 6 ISSUED**

Centurions are fast airplanes, so plan your flights around unpleasant weather or wait for the bad weather to pass.

Until you know your aircraft's performance well, based on your usual Weight and Balance and operating environment, use the POH/AFM to calculate what performance you will be expecting in climb and cruise. With the exception of actual fuel consumption — it will always be higher than the POH — I find the performance charts Cessna provides to be reasonably accurate for climb rate and cruise speed. Keep in mind that their climb rates are based on a V_Y climb, not a 120-KIAS cruise climb.

Descent planning is something many new Centurion pilots have not worked with before. This is especially true if you are flying the Turbo Centurion or the P210, as you may be flying 10,000 feet higher than the altitude you used to occupy. You have to *plan the descent!* We use the term Top of Descent, or TOD, in the planning process; it is a must-know item if you want to get down in time. TOD is a checkpoint on my flight plan form.

Most low-altitude pilots are accustomed to descending at 500 feet per minute or even less. That may not be a sufficient rate of descent if you are at FL180 and you have to descend over mountainous terrain. We can use the same techniques to slow down and then get down in all the airplanes: reduce the manifold pressure (MP) by 5 inches to slow down 20 KIAS or descend 500 FPM; then as the engine cools in two or three minutes, it may be appropriate to reduce power by another 4-5 inches MP. As long as you have an engine analyzer and are controlling your CHTs (cylinder heat temperatures) with closed cowl flaps and leaning to your reference temperatures, it should work well. If you need to come down faster, extending the landing gear will increase your rate of descent 500 FPM without an increase in speed. These are good numbers for planning purposes.

Actual performance will vary based on Weight and Balance as well as power settings and descent angles. When flying above 10,000 MSL it is normal to plan a 1,000-FPM rate of descent, so if you hear Center ask you to increase your descent rate, don't be surprised.

If you are flying a P210, you have another descent factor to consider. The pressurization system is "powered" off the engine. It is necessary to have somewhere between 18 to 22 inches of MP to keep the cabin pressure up, which you must do when descending through oxygen-required altitudes. You won't know at what MP your cabin pressure "pulls off" until you climb to altitude and test the system. If it is pulling off too early, take the airplane to your mechanic and have it fixed.

Some P210 owners have installed speed brakes to assist in their descent. When deployed, the speed brakes (really, lift interruption devices) will provide about a 500-FPM increase in descent rate, or, in level flight, a speed reduction of about 20 KIAS.

I plan my descents at 140 KIAS, transitioning into the approach at 120 KIAS.

We must constantly monitor our fuel throughout the flight in these airplanes, as sometimes the headwinds exceed our planning numbers. Too often we read of what seems like simple flights not arriving as planned.

NTSB Identification: WPR09LA214
14 CFR Part 91: General Aviation
Accident occurred Friday, April 24, 2009 in Hanford, CA
Aircraft: CESSNA T210 M, registration: N6584B
Injuries: 4 Uninjured

This is preliminary information, subject to change, and may contain errors. Any errors in this report will be corrected when the final report has been completed.

On April 24, 2009, about 1425 Pacific daylight time, a Cessna T210M, N6584B, impacted the terrain during a forced landing about five miles southeast of Hanford, California. The commercial pilot and his three passengers were not injured, but the airplane, which was owned and operated by the pilot, sustained substantial damage. The 14 Code of Federal Regulation Part 91 personal pleasure flight, which departed North Las Vegas Airport about two and one-half hours prior to the accident, was being operated in visual

> meteorological conditions. The flight was en route to Vacaville, California. No flight plan had been filed.
>
> According to the pilot, while en route to Vacaville, he noticed that the fuel quantity was getting lower than he was comfortable with, so he elected to divert into Hanford Municipal Airport to take on some additional fuel. During his descent into Hanford, the engine lost power, and he was unable to get it restarted. He therefore executed an engine-out forced landing in an open field, but during the landing roll, the airplane came in contact with the rough terrain.

Performance Numbers

The POH is the reference in all matters related to performance. Section 2 of the POH, Limitations, covers most of the airspeeds, power limitations, and so on. Definitely read this section carefully so you know the limitations. But that being said, you also have to read Section 3 (Emergency Procedures) to find the best glide speed, and Section 4 (Normal Procedures) to find things like V_Y (best rate of climb), V_X (best angle of climb), and the procedures and speeds for maximum performance takeoffs and climbs.

Section 5 of the POH (Performance) will give you the numbers needed for stall speeds in various configurations and bank angles. Section 5 is also where you will find the charts used to calculate takeoff distances and landing distances, cruise performance power settings and fuel flows, and range and endurance numbers. (I always add at least one gallon per hour fuel flow for any climb or cruise power setting.)

You will note that in Section 5 the numbers provided are based on a certain set of conditions that are spelled out in each performance category. The takeoff distances are related to takeoff weight, temperature, and altitude; the landing distances are based on landing at or near gross weight (something we rarely do) with the power at idle and full flaps (something a lot of pilots don't do and wonder why they used so much runway compared to what "the book" said).

Realistic cruising speeds of 165 to 180 KTAS are achieved in these aircraft, based on the power/fuel flow you apply to the engine.

Speed abbreviations used:

V_{ne}	Never exceed speed (red line)
V_{no}	Maximum structural cruising speed (top of green arc)
V_a	Maneuvering speed given at different weights
V_{fe}	Maximum flap extension speeds for different flap settings
V_{lo}	Maximum landing gear operating speed
V_{le}	Maximum landing gear extended speed
V_s or V_{s1}	Stall speed clean (bottom of green arc)
V_{so}	Stall speed in landing configuration, full flaps, power off, at most forward CG (bottom of white arc)
V_x	Best angle of climb
V_y	Best rate of climb
V_{glide}	Best glide speed based on weight (see Section 2 of POH)

Determining and then using "standard" power settings will greatly reduce your workload in the airplane, as you will know what performance to expect from the airplane during any phase of flight.

N210XX	Flight Condition	Pitch	IAS	RPM	MP	FF	TIT	Performance
Vx Climb	G↑ F↑	+13°	85	2700	36.5"	28.5	1450	↑↑↑↑↑ 800+
Vy Climb	G↑ F↑	+10°	100	2700	36.5"	28.5	1450	↑↑↑↑↑ 1000+
Cruise Climb	G↑ F↑	+4°	120	2500	31.0"	26.5	1450	↑↑↑↑↑ 700+
Cruise (75%)	G↑ F↑	+0°	140	2500	29.0"	18.6	1550	→→→→→
Vma	G↑ F↑	+1°	120	Full	20.0"	12.5	1500	→→→→→
En route Descent	G↑ F↑	-2.5°	140	2500	24.0"	13.5	1500	↓↓↓↓↓ -500
Quick Letdown	G↓ F↑	-5°	160	2500	20.0"	n/a	n/a	↓↓↓↓↓ -1000
Holding	G↑ F↑	+1°	120	2500	20.0"	12.5	1500	→→→→→
ILS Approach	G↓ F↑	-2.5°	120	Full	17.0"	n/a	n/a	↓↓↓↓↓ -640
Downwind VFR	G↓ F↑	+2.5°	100	Full	20.0"	n/a	n/a	→→→→→
Shortfield	G↓ F 30↓	As Req.	72	Full	n/a	n/a	n/a	↓↓↓↓↓ -950

Example of a performance chart for a P210N

Taxiing the Centurion

With such a large CG range, the way the airplane is loaded has a direct effect on taxiing techniques. If you have a center or aft CG, the nosewheel moves easily during taxi. However, most of the time these airplanes are not flown with all the seats full and at gross weight. Therefore, they usually have a fairly forward CG.

To taxi easily, you need to unload the weight from the nosewheel. Do this by pulling the yoke back so that the airplane can turn normally, using the rudder pedals which are connected to the nosewheel, and not using brakes unless necessary.

You will find you need to push *down* on the rudder pedal, not just forward, to maximize rudder movement and nose steering. When you push forward on the rudder pedals, you are usually pushing on the brakes. In normal taxi, you should rarely need brakes.

If you aren't turning, you should be able to relax the elevator to a neutral position—but pull it all the way back before starting the next turn. Check the position of your feet, as you may have them too high up on the rudder pedals and you will find you are putting on the brake when you think you are pushing on the rudder/steering link.

An aid to taxiing the airplane is to have the rudder trim in the takeoff position—that is, about half an inch or more to the right of the center mark on the rudder trim indicator (this assumes your rudder trim and rudder trim indicator are rigged correctly).

It is important that the nosewheel is centered before the run-up. If it is cocked to one side, as you go to leave the run-up area, it may take extraordinary rudder effort and use of heavy braking to turn the airplane in the correct direction.

Run-Up

The use of a checklist specifically made for the airplane as it is equipped is a key to ensuring your airplane is ready for flight.

Checklist attached to the yoke with Velcro

We have all heard about the dangers of shock cooling an engine in flight. But the engine builder RAM Aircraft is much more concerned about heat shock than shock cooling. This is such an important topic that you will find their article "Engine Shock Heating During Takeoff" in Appendix A of this book.

Many of us have heaters on our engines for use in cold weather. I have a TANIS unit that I plug into an extension cord. On even the coldest days it keeps the CHTs and oil above 60° F. The engine heater keeps me from having to sit and idle at 900-1,000 RPM for ten or more minutes to warm up the engine on cold days.

Here are the basics. Don't do an engine run-up until your CHTs are all at 200° F or greater and the oil temperature is at least 100° F. These engines need time to slowly heat up and for the oil to thin and circulate before a lot of power is applied. Also, the dissimilar metals in the engine heat at different rates. Start the engine and idle at 900-1,000 RPM as you warm the engine.

The modern engine analyzer is again necessary here, as Cessna only installed a single CHT gauge, and you will see that the cylinders don't heat up equally. Interpreting the location of 100° F on the factory oil temp gauge is not an easy task, either.

Even if the OAT is hot, you have to give the engine time to slowly heat up to the recommended temperatures before the run-up. Then, on takeoff, the application of power needs to be slow and smooth. In the case of the turbocharged engines, I bring the MP up slowly until I hear the turbocharger starting to spool up, then I pause the throttle advancement for a few seconds to let the engine stabilize before I advance the throttle to takeoff power.

Intercoolers

A number of the turbocharged Centurions have had aftermarket intercoolers installed on their engines. The name brand intercoolers I have seen installed on these airplanes are: Riley, Aircraftsman, and TurboPlus.

The intercooler is essentially a radiator that cools the induction air going into the engine. The cooler air increases the actual manifold pressure as well as provides cooler air to the engine (colder air is more dense than warmer air).

Some of the installations have a temperature differential gauge installed so that you can see the efficiency of the intercooler. Unfortunately, some do not or they were inadvertently removed.

The POH Supplement for the intercooler advises you on how to set the manifold pressure to attain the true manifold pressure (indicated manifold pressure adjusted for the manifold pressure increase provided by the intercooler) by using the temperature differential gauge

Intercooler installed on a P210N

indication. Typically, this amounts to 1 inch of additional manifold pressure for each -15° C of temperature differential. I have seen temperature differentials of up to -60° C, which equates to 4 inches of additional manifold pressure.

Many of the installations are different, even from the same manufacturer. It is imperative that you read your POH supplement and understand that you may have to have a

TurboPlus intercooler installation visible behind the fairing

lower indicated manifold pressure on takeoff that what the original POH shows you in Sections 4 and 5. Example: the redline manifold pressure is 36.5 inches and the intercooler differential gauge shows -30° C. You are supposed to reduce the indicated manifold pressure by 2 inches on the manifold pressure gauge; otherwise, you will be putting 2 inches more MP on the engine than what the manifold pressure gauge is indicating, over-boosting the engine.

Another huge factor with intercooler installations is the setup of the fuel flow. The fuel flow may not be set up to provide more than the amount of fuel shown in the POH (although on some installations I have seen it has been). The point is, if you are putting out 1-3 inches more MP than you think and the engine is not getting the corresponding increase in fuel flow, you are going to unknowingly over-temp your engine. This is another reason to have a digital engine analyzer installed if you don't already have one.

Since the intercooler is essentially a radiator, its efficiency changes as the aircraft pitch and speed changes the airflow into it, and as it moves through air of different temperatures and moisture content.

It is not uncommon to be flying along in cruise and have the manifold pressure moving up and down as much as 1 to 1 ½ inches. As long as the fuel flow movement corresponds to the MP movement, this probably does not indicate a problem.

If you have an intercooler installed, please ensure you read the POH supplement, have all of the required components installed and indicating correctly, and really understand how you are supposed to operate the engine as modified.

7

Takeoffs

Normal, Crosswind, Short, Soft/Rough

Normal Takeoff

What is a normal takeoff? The FAA *Airplane Flying Handbook* (FAA-H-8083-3A) defines a normal takeoff as: "One in which the airplane is headed into the wind, or the wind is very light. Also, the takeoff surface is firm and of sufficient length to permit the airplane to gradually accelerate to normal lift-off and climb-out speed, and there are no obstructions along the takeoff path."

Normal takeoff in a 210

Some pilots have trouble getting nice, smooth takeoffs in the 210. Some folks hold it on the runway too long (usually pilots who have been flying twins), and others try to tug it off the runway early like you can a C172. Neither of these methods will yield consistently favorable results. There are a couple of reasons why these techniques can lead to incidents or accidents, and they can be read about repeatedly in the AOPA Air Safety Foundation Safety Review of these aircraft.

After we review what the POH says we should do, we will review takeoff techniques that I have found work well for most pilots.

An excerpt from the POH:

> **NORMAL TAKEOFF**
>
> (1) Wings Flaps — 0° to 10° (10° preferred)
> (2) Power — (the setting varies with the aircraft model)
> (3) Mixture — (whatever the specific POH calls for)
> (4) Elevator Control — lift nose at 60 to 70 KIAS (for the T210L)
> (5) Climb Speed — 80-90 KIAS (based on the model)
> (6) Brakes — APPLY momentarily when airborne
> (7) Landing Gear — RETRACT in climb out
> (8) Wing Flaps — RETRACT

Please note that there is *not* a specific speed at which you are supposed to pull the nosewheel off the ground.

Some pilots, especially those who fly twins and jets, use a specific rotation speed (V_r). They hold the airplane on the runway until they see that specific IAS, and then they pull the nose up to some predesignated pitch attitude. Since Cessna does not anywhere specify an exact takeoff pitch attitude, you have no means of determining what pitch attitude to use if you are new to the airplane. Holding it on the runway too long can cause maintenance issues on the nose gear and may result in an off-runway excursion during takeoff as the aircraft "wheelbarrows" on the nosewheel.

Pilots unaccustomed to flying heavy, complex aircraft have a tendency to lift the nose to a very high nose-up pitch attitude at liftoff, regardless of the indicated airspeed — they try to pull the airplane off the ground. That doesn't work well, either.

There is no specific rotation speed for a Centurion! This aircraft has very large CG and weight ranges. When flying twins and jets, we can calculate a V_r or the POH gives us a specific number. Centurion pilots need to realize that they don't know what speed at which the aircraft nose should be rotated, as we don't have the charts to calculate a V_r, nor is such a speed provided. We don't need to calculate that speed, anyway. What we need to learn is good techniques for takeoff in various flight conditions, weight, and CG.

Setting up for the takeoff — items to prepare and know ahead of time:

- An accurate Weight and Balance: how much does the aircraft really weigh and where is the CG?
- Elevator trim setting (based on CG)
- Rudder trim setting (always to the right of center if rigged properly)
- Flaps — set to 10° (Taking off with less than 10° of flaps will dramatically increase your takeoff roll distance and make for a less positive takeoff experience.)
- Available power based on density altitude
- Field elevation and temperature
- Crosswind component

Normal takeoff procedure I use:

(1) Wings Flaps — set to 10°.
(2) Power — Smoothly apply full takeoff power and verify on the engine instruments.
(3) Mixture — In a turbocharged engine, always use full rich for takeoff. In a normally aspirated airplane at high density altitudes, I lean to peak and then increase mixture to attain 100° F rich of peak per the POH.
(4) Elevator Control — Hold the elevator in a neutral position from the beginning of the takeoff roll; the yoke will be about half-way back from the instrument panel. (This decreases drag and prevents "pumping" of the nose strut as lift is

generated.) As the aircraft starts to get light on the ground (depending on the airplane, somewhere between 50 and 60 KIAS), apply enough back pressure to *gently* lift the nosewheel 3-5 inches off the ground and wait. If you do this correctly, when the airplane has sufficient speed it will just fly off the runway, regardless of weight. The key here is to raise the pitch attitude only +4° to +5° (with the nosewheel just off the ground) and hold that pitch attitude. The nose will try to rise as the airspeed increases, but you want to maintain that +4° to +5° pitch attitude after you lift off. If you do, the speed will quickly increase to climb speed. Unless I am performing an obstacle takeoff, I never raise the nose above +7° pitch.

(5) Climb Speed — 80-90 KIAS (based on the model) and then accelerate to V_Y or cruise climb speed.

(6) Brakes — APPLY momentarily when airborne. This doesn't mean stand on the brake pedals; it means lightly tap them with your toes, just enough to stop the wheel rotation. Remember, the wheels are airborne now and already slowing to a stop on their own.

(7) Landing Gear — RETRACT in climb-out. (If properly trimmed for takeoff, you will need to put in some nose-down trim as the gear comes up and the speed continues to increase.)

(8) Wing Flaps — Once airspeed is up to 85 KIAS or greater, RETRACT and trim as necessary for the desired climb angle. If you retract the flaps early (at a low airspeed), the plane will sink immediately — and you will know it! That sink is especially noticeable when the airplane is near gross weight or if you retract the flaps just as you come out of ground effect at a low airspeed. Note: when the flaps are retracted, the nose will want to come down, so you will have to re-trim.

The pitch attitude at and immediately after liftoff is the key to consistently having nice, smooth takeoffs. Once the speed has increased to

80-85 KIAS, the airplane will start climbing. You will want to retract the landing gear when there is no usable runway remaining or when you are high enough above the runway that you could lower the gear again if you had an engine failure on climb-out. Don't retract the gear early or when very close to the ground, because it drops down about 24 inches lower than when in the down-and-locked position during the retraction cycle.

210 with gear in transit

Pitch attitude for a V_Y climb is typically +7° at takeoff power, +12° for a V_X climb if you need best angle of climb for some reason, and +4° for a 120-KIAS cruise climb. Note: the pitch attitudes mentioned here are based on an attitude indicator that has been set correctly in flight. Some aircraft attitude indicators' "pipit" is not set correctly or has been moved by pilots or technicians. Once you are sure your pipit is set correctly, then the pitch attitudes I mention should work for you.

Now that you have some parameters to work with, you need to go out and practice and perfect these techniques with a qualified C210 instructor on board in day VFR conditions, until you can automatically perform good, consistent takeoffs without thinking about it.

7° pitch-up attitude (pipit indicated)

4° pitch-up attitude

A note about landing gear retraction times and effects. When you retract the landing gear, the airplane will yaw and wallow a bit. The airplane does somewhat of an entire body wiggle. This is caused by the gear doors opening in some aircraft and the gear dropping down and turning sideways into the airflow on all 210s. Then, of course, the gear struggles up into the rear of the airplane like an uncoordinated stork, with one gear usually locking in place before the other. If you have gear doors, when the gear doors close there is additional movement of the airplane.

Gear cycle times are important to you, the pilot. Make note of and remember the time it takes for your landing gear to completely cycle up, as well as the time it takes to extend. These times will probably be different, and if the time it takes to cycle the gear in either direction changes, it may indicate a mechanical or electrical problem with the gear system. The landing gear fluid reservoir is cleverly hidden on the right side of the center pedestal, about where the co-pilot's left ankle will be found. This is a "must-check" item during the preflight.

The 210 has a history of the gear pump continuing to run after the gear has been cycled, so be alert for the pump sound to stop and the ammeter to no longer show a discharge when operating the gear. If it doesn't shut off in the normal time, you should refer to Section 3, Emergency Procedures, where among other things it is suggested

The wheels are extended further from the fuselage while in transit than when in the down-and-locked position

you pull the gear pump circuit breaker. Take note that once you pull that circuit breaker, moving the gear control handle will not cause the gear to fully extend or retract. See Section 3 of the POH/AFM before attempting to land. Some pilots have actually had a gear pump "on" light installed to help them recognize the situation if the pump continues to run.

You also need to know how long it takes for the gear to go down and lock from the time you put the gear handle down until you see a green light on the panel. If you have an emergency and are trying to reach a suitable landing place, you may need to keep the gear up as long as practical and then extend it at the last possible moment.

Obviously, the reason to know both cycle times is in the event you have an engine failure on takeoff. It makes sense not to retract the gear with runway remaining if you won't have time to put it back down should you lose the engine. Knowing that down-cycle time allowed me to reach a runway with the gear up for maximum glide and then extend the gear seconds before touchdown following an engine shutdown.

Crosswind Takeoff

The POH does not specifically address crosswind takeoffs and there probably isn't any reason it should. The C210 series aircraft handle crosswinds very well. All you have to do is apply the same techniques you learned as a private pilot. You take off with the normal takeoff flap setting, applying the same techniques you would in any other tricycle gear airplane.

Short Field Takeoff

Your POH tells you exactly how Cessna wants you to perform this takeoff. Again, here is an excerpt from a POH, this one a 1979 P210N:

(1) Wing Flaps — 10°
(2) Brakes — Apply
(3) Power — 36.5 INCHES Hg and 2,700 RPM (5 minute limitation)
(4) Mixture — Adjust to 180 lbs/hr
(5) Brakes — RELEASE
(6) Elevator Control — LIFT NOSE WHEEL at 65 KIAS.
(7) Climb Speed — 78 KIAS until all obstacles are cleared
(8) Landing Gear — RETRACT after obstacles are cleared
(9) Wing Flaps — RETRACT after reaching 85 KIAS

NOTE

Do not reduce power until wing flaps and landing gear have been retracted.

Remember that the landing gear extends at least two feet lower during the retraction cycle. Do not retract before clearing the obstacles as the POH/AFM states!

Short-field takeoff procedure I use:

(Caution: I am an experienced commercial pilot who has thousands of hours in these airplanes and flies them regularly in a variety of challenging environments. This is a technique that those of you with a commercial rating may recognize; however, I am not recommending you use this technique. I am merely relating to you a technique that I have used in certain actual short-field operations. You should operate the aircraft in accordance with the Approved Flight Manual, POH, or Owner's Manual and the appropriate Flight Manual Supplements specific to your aircraft.)

(1) Wing Flaps — 10° (unless the aircraft has STOL modifications)
(2) Position the aircraft so that every inch of available runway is usable. I refer to this as "hanging the tail off the end of the runway." I want every bit of runway I can get and I am not leaving any behind me.
(3) Offset the nose to an angle of approximately 10° to the right of the runway centerline. (I can't keep it from turning left when the brakes are released at full static power, so this technique allows for the airplane to move left, back to the centerline at brake release without stomping on the brake and inhibiting acceleration in a vain attempt to hold centerline.)
(4) Brakes — APPLY. *Stand on them! Don't let the airplane move an inch until you want it to.*
(5) Power — FULL POWER smoothly applied. MP to Redline. The prop will not turn 2,700 RPM when static, so I don't expect it; instead, it will be closer to 2,600-2,650. The prop will go to 2,700-RPM redline once we have about 60 KIAS.
(6) Mixture — Full rich mixture on a turbocharged engine. I am looking for the fuel flow to be near redline or maybe even a little beyond. On a normally aspirated 210, I will adjust to 125° rich of peak for best power, anticipating more power once the prop has enough air flow to go to redline.

(7) Brakes — RELEASE and move the feet down to the rudder pedals.
(8) Elevator Control — Neutral elevator until 65 KIAS if the plane is light, 70 if near gross weight. Then a steady, smooth pull on the yoke to raise the nose to +12° without having the stall warning horn come on. If the horn comes on, gently lower the nose until it goes quiet.
(9) Climb Speed — 78 KIAS until all obstacles are cleared.
(10) When clear of obstacles, lower nose to V_Y or cruise climb attitude and let the speed start building.
(11) Landing Gear — RETRACT after reaching V_X.
(12) Wing Flaps — RETRACT after reaching 85 KIAS.
(13) Do not reduce power until at least 1,000 feet AGL.

For a maximum performance takeoff, the POH requires 10° of flaps.

The stock C210 outperforms both the T210 and P210 at lower altitudes and cooler temperatures. At low altitudes and in cool weather, I think it is a great short-field performer, but I find the C210 short-field takeoff procedure severely affected in high density altitude situations, primarily because the engine cannot develop 100% horsepower — and you can't really calculate how much horsepower/performance you are going to get. Use an abundance of caution in short-field operations during very high density altitude conditions in the C210.

The T210 has very good short-field performance when flown properly, especially if it is a few hundred pounds under gross weight. I like to take the "book" ground roll or obstacle clearance numbers found in Section 5 of the POH/AFM and add 50% to them as an insurance factor. Remember that if you aren't climbing out *over* an obstacle, the ground run is the key number, not the distance over an obstacle that isn't there. Obviously, if there is an obstacle, caution trumps bravado.

The P210 is not much of a short-field performer compared to the other two models and is known for its slow climb rates. I don't recommend short-field operations in P210s unless you are highly skilled and really know your aircraft and the short fields in which you intend to operate.

TAKEOFF DISTANCE
3700 LBS AND 3400 LBS
SHORT FIELD

REFER TO SHEET 1 FOR APPROPRIATE CONDITIONS AND NOTES.

WEIGHT LBS	TAKEOFF SPEED KIAS		PRESS ALT FT	0°C		10°C		20°C		30°C		40°C	
	LIFT OFF	AT 50 FT		GRND ROLL	TOTAL TO CLEAR 50 FT OBS	GRND ROLL	TOTAL TO CLEAR 50 FT OBS	GRND ROLL	TOTAL TO CLEAR 50 FT OBS	GRND ROLL	TOTAL TO CLEAR 50 FT OBS	GRND ROLL	TOTAL TO CLEAR 50 FT OBS
3700	69	75	S.L.	950	1565	1035	1705	1125	1865	1230	2045	1340	2245
			1000	1010	1655	1100	1805	1200	1975	1310	2170	1430	2385
			2000	1075	1750	1170	1915	1275	2095	1395	2300	1525	2535
			3000	1145	1855	1245	2030	1360	2225	1485	2445	1625	2700
			4000	1215	1965	1330	2155	1450	2365	1585	2605	1735	2880
			5000	1300	2085	1420	2290	1550	2515	1695	2775	1855	3070
			6000	1385	2215	1515	2435	1655	2675	1810	2955	1980	3280
			7000	1480	2355	1615	2590	1770	2855	1935	3155	2120	3510
			8000	1580	2505	1725	2760	1890	3045	2070	3370	2270	3755
3400	66	72	S.L.	780	1290	850	1405	925	1530	1010	1670	1100	1830
			1000	830	1365	905	1485	985	1620	1075	1770	1170	1940
			2000	885	1440	965	1570	1050	1715	1145	1875	1250	2060
			3000	940	1525	1025	1665	1120	1820	1220	1990	1330	2185
			4000	1000	1615	1095	1765	1190	1930	1300	2115	1420	2325
			5000	1070	1715	1165	1870	1270	2050	1390	2250	1515	2475
			6000	1140	1815	1245	1985	1355	2180	1480	2390	1620	2635
			7000	1215	1930	1325	2110	1450	2315	1585	2550	1730	2815
			8000	1300	2050	1415	2245	1550	2465	1695	2715	1855	3005

Figure 5-4. Takeoff Distance (Sheet 2 of 2)

That being said, my definition of a short field is a landing surface less than 2,500 feet long for the P210.

An exception to this caution is the P210 Silver Eagle, which has been modified with the installation of a Rolls-Royce (Allison) 250-B17F/2 gas turbine engine. This turboprop engine puts out 450 SHP under perfect conditions—you won't need much runway to be airborne. This conversion makes the P210 a totally different airplane, though I won't get into discussing it at this point.

If you want to operate the Centurion aircraft routinely out of short fields or soft fields, a STOL modification is the way to go! The performance of all C210s is dramatically improved with STOL modifications, as mentioned in the previous chapter.

I believe the STOL modifications make them a much safer airplane to operate in all environments. The modifications are simple: a leading edge cuff, different wing tips, and a stall fence on the top of each wing.

However, if you want or need anti-ice capability, you can't have the STOL modification. They are mutually exclusive on the Centurion.

My own T210 has a Horton STOL modification and I love the advantages it has given me in slow speed maneuvering, shorter takeoff and

STOL leading edge cuff being installed on a P210N

P210N with TKS anti-ice and VGs

landing distances, etc. There are no moving parts, lower stall speeds, greater control at low speeds, and *superior* short- and soft-field operations! Should you own a C210 with a STOL modification, or if you have STOL installed, find an instructor that can teach you how to fly the airplane in STOL mode. You won't believe the shortened takeoff runs, low-speed maneuverability, and short-field landings you can safely perform. If you haven't been in a STOL 210 when it takes off with 30 degrees of flaps, you probably wouldn't believe it is able to takeoff with the flaps that far extended. But it does, and it does it in such a short distance it amazes you. (This is *not* an obstacle departure takeoff.) The first time you take off in 500 feet or less, you won't believe it happened!

VGs work very well on all the C210s, as they enhance the short/soft-field capability, improve low-speed handling, and lower the stall speeds. Teaching pilots how to fly their aircraft with STOL installed is one of my great pleasures as a 210 flight instructor.

Soft/Rough Field Takeoff

Cessna does not address this topic in their POHs and I can't really blame them. The variety of environments you can experience with "soft fields" is boundless.

Grass fields can have grass that is short or long, wet or dry, and anything in between. Sand can be soft, hard, wet, or a combination of all of those things on the same "runway," which probably isn't really a runway at all. Dirt roads, open fields, rough fields, beaches, gravel, etc. were all used for runways at one time or another. Under the right circumstances, the C210 aircraft can fly on and off them.

T210 at Sulphur Creek, Idaho

Soft field operations are taught to private pilots and tested for during the practical test. Commercial pilot applicants must also demonstrate knowledge and skill with soft field takeoffs and landings (they are ignored for the ATP rating).

Pilots have been flying C210s off soft fields since they started making them over forty years ago. Remember the 1963 photo of one on a polo field in Chapter 1? The older 210s have better soft/rough field performance than the newer and heavier models. As with any maximum-performance operation, knowledge and skill are the keys to safe operations of this kind.

If you want to fly off a variety of runways, it takes a little studying for some basic knowledge. Start with the FAA's *Airplane Flying Handbook* again (FAA-H-8083-3A). Chapter 5 covers takeoffs and Chapter 8 discusses landings. Read it thoroughly. Digest the information. Then read it again and make 3x5 cards for each technique. Review Chapter 5 of your POH thoroughly. Then, find that *very* experienced C210 instructor. After your ground school portion is over and you have written out and briefed your first flight to soft/rough fields, go out and start enhancing your skill. Have some fun in a wonderfully capable airplane.

This dry lake, almost a mile long, is a favorite of many pilots

Thoroughly brief the techniques to be used and understand them completely before you even preflight the airplane. There shouldn't be questions about technique before you take flight. Sure, you will have questions about how it will feel and what it will do, but the fundamentals of the soft/rough field techniques that you intend to apply should be completely understood. Just learning how to "read a landing area" is a challenge, and you may have no experience in it at all. That is where the instructor's experience comes into the equation. If the instructor has never landed on a dry lake in a C210 and you intend to do it, find an instructor that has that experience. From my own experience, I can tell you that what may look like a dry lake may not be "dry" at all—but there are ways to figure out whether it is safe for landing or not.

My T210 has taken off from and landed on dry lakes, beaches, gravel and dirt roads, open fields, grass runways, sand "runways," and a variety of all of the above. I've had no damaged props, no damage at all to the airplane—ever. Careful analysis of the intended landing site, knowledge and experience in flying the airplane, and application of the techniques found in the FAA's *Airplane Flying Handbook* work!

The off-airport utility of the Cessna 210 has provided me with the opportunity to visit many wonderful and exciting places that I could not have otherwise reached, including portions of Alaska, Canada, Mexico, Greenland, Iceland, the UK, and western Europe. Even the heavy P210 can routinely make use of many of these types of landing areas when the surface is known and if the airplane is flown properly.

Refueling in Greenland

A 210D in cruise (note the wing struts)

8

Climb, Cruise, Descent

Flying

Climbing the 210

The Cessna 210 and T210 have very nice "real world" climb performance, but you will find the P210N climbs a bit slower. The P210N has adequate climb performance in standard temperature environments, but it takes careful operation in hot climbs. If you live or fly in hot climates, you may find high oil temperatures and CHTs during high ambient air temperature climbs with the T and P models. The R model aircraft are not as affected by hot ambient air operations. These airplanes have big engines that need a lot of air and fuel to help cool them in the climb during high temperature operations. Think of it this way: the airflow cools the outside of the engine, the fuel cools the inside.

Don't be surprised if your airplane doesn't make the rate of climb number in the POH, as they are based on a standard weight aircraft that is brand new (minimally equipped) and flown by a factory test pilot.

Your airplane probably weighs more than the flight test aircraft that was used to develop the numbers in the chart above. Your aircraft probably has more drag from optional equipment (antennas, air conditioning, scoops, etc.) and doesn't have a brand new engine and propeller. So if you are using the same climb configuration as shown on the climb rate chart, yet getting 50-100 FPM less than the factory climb rate, don't be disappointed by that — but remember it and use it for your flight planning purposes. This kind of information is critical to you when calculating minimum climb gradients on IFR departures. Use "real" information when you know it, not just published data.

RATE OF CLIMB

MAXIMUM

CONDITIONS:
Flaps Up
Gear Up
2600 RPM
Cowl Flaps Open

PRESS ALT	MP	PPH
S.L. to 17,000	35	162
18,000	34	156
20,000	32	144
22,000	30	132
24,000	28	120

WEIGHT LBS	PRESS ALT FT	CLIMB SPEED KIAS	RATE OF CLIMB - FPM			
			-20°C	0°C	20°C	40°C
4000	S.L.	100	1170	1035	895	755
	4000	100	1080	940	800	655
	8000	100	980	840	695	555
	12,000	100	870	730	590	- - -
	16,000	100	740	605	470	- - -
	20,000	99	485	355	- - -	- - -
	24,000	97	190	70	- - -	- - -
3700	S.L.	99	1310	1165	1020	875
	4000	99	1215	1070	925	775
	8000	99	1115	965	815	670
	12,000	99	1000	855	710	- - -
	16,000	99	865	730	590	- - -
	20,000	97	600	470	- - -	- - -
	24,000	95	295	170	- - -	- - -
3400	S.L.	97	1465	1320	1165	1015
	4000	97	1370	1220	1065	910
	8000	97	1265	1110	955	795
	12,000	97	1150	995	845	- - -
	16,000	97	1010	865	725	- - -
	20,000	96	730	595	- - -	- - -
	24,000	94	405	275	- - -	- - -

Figure 5-5. Rate of Climb

As you already know, when you climb with a normally aspirated engine, the higher you climb, the less manifold pressure you can maintain. Thus, in the Cessna 210, your available horsepower decreases as you climb and your rate of climb will diminish accordingly, even if at full throttle and maximum-allowable RPM. You will of course be leaning the mixture in the climb to maintain best available power, using the Cessna Power Computer or the "Maximum Performance Take-Off and Climb Setting" placard that is located adjacent to the fuel flow indicator.

Manifold pressure and fuel flow indicator

Those of you that own or fly Cessna 210s with the Continental IO-550 engine conversion will love the increased performance and the fuel control system; it has an altitude-compensating fuel system for climb. You still may have to manually lean for cruise, depending on the power setting you are using. When the IO-550 is installed, an engine manual is included. It is well worth your time to read how that engine operates, as with any modification, your original POH will not reflect how to manage the IO-550.

The turbocharged aircraft can normally attain sea-level horsepower up to 18,000 feet, so you can set a desired manifold pressure and RPM and climb at that power and rate of climb to the flight levels on a standard day.

A word of caution for the turbocharged aircraft operators. If you lean to the mixture settings shown in Section 5 of the POH, you will be running the engine with a very lean fuel/air ratio. This will result in high engine temperatures and shorter engine life. Engine temperature limits (oil; CHT; exhaust gas temperature, or EGT; and turbine inlet temperature, or TIT) are presented with the maximum temperatures, but that doesn't mean you can run the engine all day long with the temperatures at redline. Most 210 operators keep the CHT at less than 400° (preferably in the 380° range or less) and the TIT in the vicinity or 1500°-1600°, depending on the other temperatures. The real key to operating these engines is keeping them cool. If you operate with oil below 200°, CHTs in the 380° range, and EGT/TIT wherever you need it to be to keep the CHTs in the desired range and the oil cool, you should be able to take your engine very close to TBO.

When operating in high ambient air temperature conditions, we must use all available techniques to keep these engines cool. These techniques include, when necessary: climbing with full-rich mixture all the way to cruise altitude, fully opening cowl flaps during the climb, and using a lower angle of attack/higher airspeed than V_Y. (I cruise climb most 210s at 120 KIAS to accomplish two things: one, keeping the engine cool; two, allowing me to see over the dash and look for traffic.) Also, although RPM is primary power, it may be necessary to use a lower RPM and a higher MP to maintain the same horsepower output, since a lower RPM generally will yield lower CHTs.

P210 that has transitioned to a 4° cruise climb during a missed approach

Sometimes it is necessary during climb to use the high boost setting on the fuel pump during climb to keep the CHTs in the desired range. To do this, you have to hold the red auxiliary fuel pump switch in the ON position, as it is spring loaded. When you no longer need the high boost pump running and you release the red switch, you must turn off the yellow auxiliary pump switch, as it is not spring loaded. You must be careful if you use this technique, because if you forget to turn off the yellow switch, the engine may die when you retard the throttle. You can learn more about this yellow switch in Chapters 3 and 7 of your POH/AFM.

The Engine Analyzer — what Cessna didn't have to include when the airplanes were built — is now a crucial part of our engine operations in the Centurion product line. We are able to accurately see all of our engine parameters in a single instrument if we have one installed. There are two major companies that produce engine analyzers normally installed in 210s: JPI Instruments and Electronics International. They both provide the desired data in a very usable manner.

85% power climb

In this book I am going to be referring to the JPI products and using pictures of them in flight because that is what is in my airplanes, but both companies' products provide the information if you are looking to install one in your airplane. You should evaluate the various products and then install the one you prefer.

Cruising

You will find that when you go to level off at your assigned/desired altitude in a 210, you seem to level off low every time. If you were taught to start the level off at an altitude equal to 10% of the rate of climb, you will never get to the desired altitude. This airplane needs to be climbed right up to the desired altitude and then have the nose slowly lowered to allow the speed to build; otherwise, you will descend back down 50-100 feet. If you are in a cruise climb at 4° pitch, as you level off you should first lower the nose to around 2 ½° and let the speed build on the airplane while you maintain altitude. As the airspeed builds to cruise, you will then be able to lower the nose to the horizon and maintain altitude.

As you are leveling off, you will be feeding in nose-down trim and left rudder trim to adjust the airplane for hands-off, straight-ahead level flight. If the airplane is properly rigged, you will find that once you trim it up, it flies straight and true and maintains the altitude in smooth air.

Now we move on to that essential "cruise" portion of the checklist. It is time to reduce power to the cruise setting we desire. (Any power change will result in the necessity of re-trimming the airplane.) As with all airplanes with controllable pitch props, and especially if turbocharged, we *reduce power* using a cockpit flow from *left to right*: reduce manifold pressure, reduce propeller RPM, lean the mixture, adjust cowl flaps, re-trim.

When we increase power from cruise—for instance, to climb to a higher desired/assigned altitude—we *increase power* with a cockpit flow from *right to left*: first we open the cowl flaps, increase the mixture

CRUISE PERFORMANCE

PRESSURE ALTITUDE 8000 FEET

CONDITIONS:
4000 Pounds
Recommended Lean Mixture
Cowl Flaps Closed

NOTE
For best fuel economy at 70% power or less, operate at 6 PPH leaner than shown in this chart or at peak EGT if an EGT indicator is installed.

		20°C BELOW STANDARD TEMP -21°C			STANDARD TEMPERATURE -1°C			20°C ABOVE STANDARD TEMP 19°C		
RPM	MP	% BHP	KTAS	PPH	% BHP	KTAS	PPH	% BHP	KTAS	PPH
2500	30	- - -	- - -	- - -	81	177	107	76	176	100
	28	80	172	105	75	172	99	71	170	93
	26	74	167	97	70	166	91	65	164	86
	24	68	160	89	64	159	84	60	156	79
	22	60	152	80	57	150	76	53	147	72
2400	30	81	173	106	76	173	100	72	171	94
	28	75	168	99	71	167	93	67	165	87
	26	69	162	91	65	161	86	61	158	81
	24	63	155	83	60	153	79	56	151	74
	22	57	147	76	54	145	72	51	142	68
2300	30	78	170	102	73	170	96	69	168	90
	28	72	165	95	68	164	90	64	162	84
	26	67	159	87	63	157	83	59	155	78
	24	60	152	80	57	150	76	54	147	72
	22	54	143	72	51	141	69	48	137	65
2200	30	73	166	96	69	165	91	65	163	86
	28	68	161	89	64	159	85	60	157	80
	26	63	154	82	59	152	78	55	150	74
	24	57	147	75	54	145	72	50	141	68
	22	51	138	69	48	135	65	45	131	62

Figure 5-7. Cruise Performance (Sheet 4 of 12)

CRUISE PERFORMANCE

PRESSURE ALTITUDE 16,000 FEET

CONDITIONS:
4000 Pounds
Recommended Lean Mixture
Cowl Flaps Closed

NOTE
For best fuel economy at 70% power or less, operate at 6 PPH leaner than shown in this chart or at peak EGT if an EGT indicator is installed.

		20°C BELOW STANDARD TEMP -37°C			STANDARD TEMPERATURE -17°C			20°C ABOVE STANDARD TEMP 3°C		
RPM	MP	% BHP	KTAS	PPH	% BHP	KTAS	PPH	% BHP	KTAS	PPH
2500	30	---	---	---	81	191	106	76	188	100
	28	80	186	105	76	184	99	71	182	93
	26	75	180	99	71	179	93	67	176	88
	24	70	173	91	66	171	86	62	168	81
	22	63	165	83	60	162	79	50	158	74
2400	30	81	187	107	77	186	101	72	183	94
	28	76	181	100	72	180	94	67	177	88
	26	71	175	93	67	173	88	63	170	83
	24	66	168	86	62	166	81	58	162	77
	22	60	160	79	57	157	75	53	153	71
2300	30	79	184	103	74	183	97	70	180	91
	28	74	179	97	70	177	92	66	174	86
	26	69	172	90	65	170	85	61	167	80
	24	63	165	83	59	162	79	56	158	74
	22	57	156	76	54	153	72	51	148	68
2200	28	70	174	92	66	172	87	62	169	82
	26	65	168	86	62	166	81	58	162	77
	24	60	160	79	56	157	75	53	153	71
	22	54	150	72	51	146	68	48	141	65
	20	48	139	65	45	134	62	42	128	59

Figure 5-7. Cruise Performance (Sheet 8 of 12)

(usually full rich in the turbocharged versions), increase the propeller RPM, add manifold pressure, and re-trim.

Cruising with the cowl flaps partially open is a way of life for the turbocharged versions in warm weather; the higher you fly, the more open you may have to have them. The air is much thinner at higher altitudes, and since you are still putting out somewhere between 60-75% power, opening the cowl flaps to maintain acceptable CHTs is normal. Flying with the cowl flaps open does slow you down, but going a little slower is better than overheating the engine. A nice thing about the turbocharged models is that the higher you fly them (up to a point), the faster they go, so any airspeed loss caused by the cowl flaps is gained back by having a higher true airspeed at higher altitudes.

The issue of rich-of-peak operations versus lean-of-peak engine operation is always discussed during the training courses I teach. Books, videos, even entire courses are taught concerning lean-of-peak operation in these airplanes. For the purposes of this book, I will address the issue by only discussing rich-of-peak operations, as that is the method found in the POH/AFM and in the manuals/charts provided by the engine manufacturers Continental and RAM.

I know a number of Centurion owners that routinely operate their airplanes lean of peak, and many swear by it. But before you do, I would suggest you take some quality training first. A digital engine analyzer is imperative if you are going to run lean of peak.

The basic digital engine analyzer provides us with cylinder head temperatures, exhaust gas temperatures, and the turbine inlet temperature. With a state-of-the-art engine analyzer we can see all six cylinders' temperatures, instead of just one as Cessna provided when the aircraft was manufactured. Technology has definitely helped us with our engine operations and troubleshooting.

A more advanced engine analyzer provides even more data, as shown in this image of the EDM 830: actual oil temperature and oil pressure in exact numbers, fuel flow in gallons or pounds based on the pilot's desires, as well as the CHTs, EGTs, and TIT. We even get the OAT,

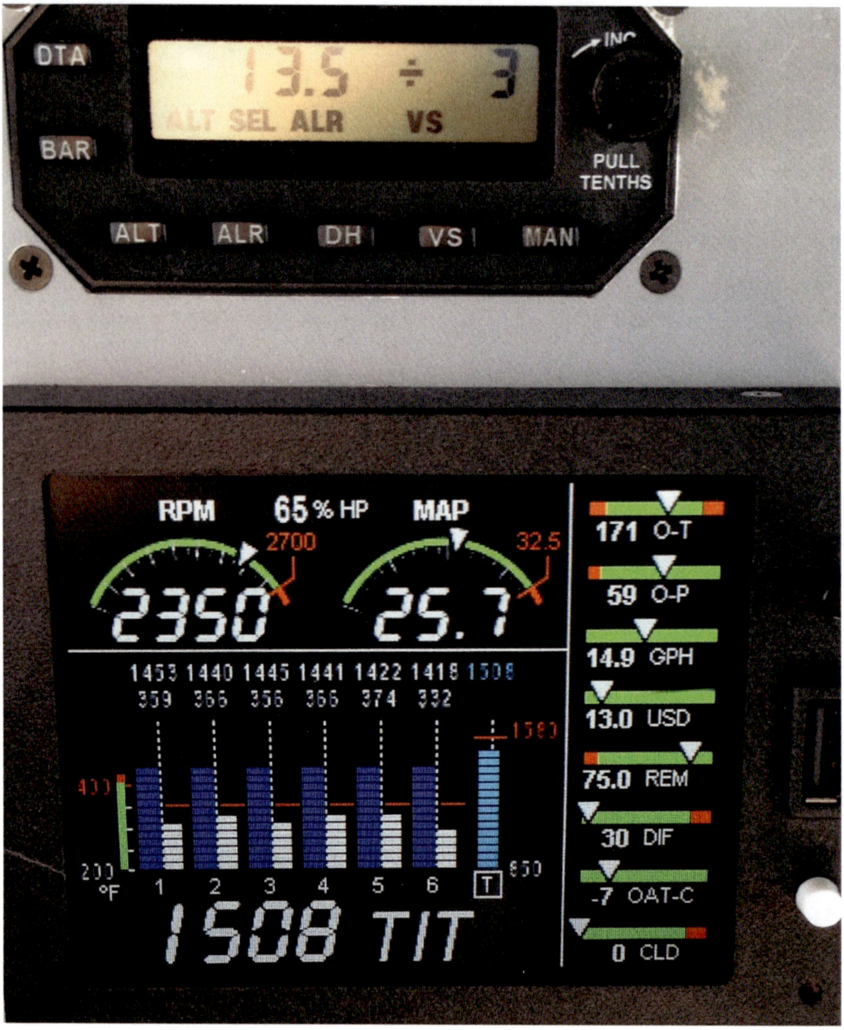

65% cruise at 13,500 feet — EDM 830

which can help us understand why the CHTs are so hot or so cool. And best of all, you can get your engine's percent horsepower output!

The most advanced analyzers are STC'd (Supplemental Type Certificate from the FAA) to replace all of our engine-related instruments, including fuel gauges, ammeter, etc. All of the old instruments can be removed from the panel, as shown in this EDM 930 installed in a P210N.

EDM 930 in flight

We have performance charts in Section 5 of the POH/AFM to assist us in planning our flight, but they may not reflect what your airplane really does. They were developed in a flight test aircraft, and you can bet it was not equipped exactly like yours.

Take Section 5 and use it as a format to create your own aircraft's performance, especially the fuel burn. There are so many variable power settings listed in the charts that you will never use them in real-life operations, so don't try to create them all.

I fly primarily using four power settings: one power setting for climb, one for cruise, one for descent, and one for when I am in a VFR traffic pattern. I do this in my airplane using the horsepower percentage. Why? Because I really don't need a whole bunch of power settings. I want to be able to analyze the performance of the aircraft in the four regimes in which I fly — climb, cruise, descent, and approach to land.

Climb	80%	32"	2500	1450 TIT	26.5+
Cruise	75%	30"	2350	1525 TIT	18.1
Cruise	65%	26"	2350	1550 TIT	15.5
Descent	55%	21"	2350	1500 TIT	12.5
Dnwnd	50%	20"	Full	#1EGT <1500	

Owner-created power settings placard

I like to go fast in my Centurion, so I normally cruise at 75% power. (If I have a nice tailwind, I can dial the power back to 70%, 65%, or even less if I desire—and I can do it using the "% HP" shown on my engine analyzer.) But maybe you don't want to burn that much fuel in cruise. Maybe you want to use a more economical power setting of 65%, which is considered by many pilots and engineers to be the best overall power setting for speed and economy. That's fine, it's your airplane. Make your chart accordingly.

I usually cruise at 75% power regardless of the altitude at which I am flying. Why? It just makes it easier because there are fewer numbers to try to digest and less tinkering with the engine. I know what each of my engine parameters should be reading, and if they aren't showing me what I expect, I look around to see why. If I am flying at a variety of different power settings, then there are no "standard" numbers, which makes it difficult to detect engine (especially temperature) anomalies.

RAM Aircraft, who supplies many of us with our engines, has performed extensive testing on the engines and has published their own recommended power settings—with a caveat that you must be using an electronic fuel measurement system. In other words, they want you to have a digital engine analyzer with fuel flow when you are running their engines.

Nov. 18, 2003
RAM PN: 2325

RAM Recommended Power Settings wh
using an Electronic Fuel Measurement System.

Page 1
Fuel is computed at 5.85 pounds/gallon.

Cessna T210 & T206 and P210 (1978 -1981) 310 hp TSIO-520-M, -R, or -P

Operation outside limits specified below and on reverse page is not approved.

Aircraft Configuration	MP Setting	RPM	Fuel Flow Setting	EGT °F ±25°F	CHT °F	Oil Temp °F
Takeoff	36.5″ MP	2700	Full Rich 30 to 31 gph 175 to 181 pph	1450°F to 1550°F	200°F min. 300°F to 350°F Ideal	100°F min. 150°F to 190°F Ideal
Climb	30″ MP	2500	24.5 gph 143 pph	1450°F Calibration Setting at 10,000 ft.	440°F max. 380°F to 410°F Ideal	230°F max. 150°F to 190°F Ideal
232 hp Cruise 75 %	Set MP: Per Altitude & OAT See Reverse Side.	2400	18.5 gph 108 pph	1650°F Maximum	440°F max. 360°F to 410°F Ideal	220°F max 150°F to 190°F Ideal
202 hp Cruise 65 %	Set MP: Per Altitude & OAT See Reverse Side.	2400	16.0 gph 93 pph	1650°F Maximum	430°F max. 360°F to 410°F Ideal	210°F max. 150°F to 190°F Ideal
202 hp Cruise	Set MP: Per Altitude & OAT See Reverse Side.	2300	16.0 gph 93 pph	1650°F Maximum	430°F max. 360°F to 410°F Ideal	210°F max. 150°F to 190°F Ideal
171 hp Range	Set MP: Per Altitude & OAT See Reverse Side.	2300	13.5 gph 80 pph	1625°F Maximum	420°F max. 360°F to 410°F Ideal	200°F max. 150°F to 190°F Ideal
Normal Descent	35% to 65% Power	Pilot's Discretion	Adjust to → Maintain EGT	1300°F min. 1400°F Ideal	200°F min. 300°F to 390°F Ideal	200°F max. 150°F to 190°F Ideal
Approach and Landing	35% to 65% Power	Pilot's Discretion	Adjust to → Maintain EGT	1200°F min. 1300°F Ideal	200°F min. 300°F to 390°F Ideal	200°F max. 150°F to 190°F Ideal

• In the event of an Emergency, use Full Power at Pilot's Discretion. • If an Aftermarket Intercooler is installed, refer to other side of card.

RAM Aircraft, Limited Partnership • 7505 Karl May Drive • Waco Regional Airport
P.O. Box 5219 • Waco, Texas 76708 • 254-752-8381 • Fax: 254-752-3307

RAM Recommended Cruise Power Settings when using an Electronic Fuel Measurement System.

Nov. 18, 2003
RAM PN: 2325 T210 & T206 and P210 (1978 -1981) 310 hp TSIO-520-M, -R, or -P Page 2

	Cruise Power:	**232 hp**	**202 hp**	**202 hp**	**171 hp**
	RPM:	2400	2400	2300	2300
	Fuel Flow:	18.5gph 108pph	16.0gph 93pph	16.0gph 93pph	13.5gph 80pph
	EGT:	1650°F max.	1650°F max.	1650°F max.	1625°F max.
10,000 ft.	**MP:**				
ISA -30°F	-7°F:	28.0″	25.0″	26.0″	22.0″
ISA°F	23°F:	29.0″	26.0″	27.0″	23.0″
ISA +30°F	53°F:	30.0″	27.0″	28.0″	24.0″
15,000 ft.	**MP:**				
ISA -30°F	-25°F:	28.5″	25.5″	26.0″	22.0″
ISA°F	5°F:	29.5″	26.5″	27.0″	23.0″
ISA +30°F	35°F:	30.5″	27.5″	28.0″	24.0″
20,000 ft.	**MP:**				
ISA -30°F	-42°F:	29.0″	25.5″	26.5″	22.5″
ISA°F	-12°F:	30.0″	26.5″	27.5″	23.5″
ISA +30°F	18°F:	31.0″	27.5″	28.5″	24.5″

Note: Setting fuel flow by the peak and enrichen method is not recommended when this chart is available and the fuel flow indicator being used is electronic.

Aftermarket Intercooler Installed: Aircraft equipped with an add-on aftermarket intercooler should consult the intercooler manufacturer's Supplemental Flight Manual, which typically instructs user that manifold pressure (MP) be reduced by -2" MP during takeoff, and up to as much as -4" MP during climb and cruise -- such being in proportion to the additional temperature drop of the induction air after it crosses through the intercooler.
Manifold Pressures on this RAM Recommended Power Settings Chart (both sides) are for *non-intercooled* engines only. Using higher than appropriate manifold pressures is not approved by RAM.

RAM

RAM Aircraft, Limited Partnership •7505 Karl May Drive •Waco Regional Airport
P.O. Box 5219 •Waco, Texas 76708 •254-752-8381 •Fax: 254-752-3307

Descent

I like to descend at no more than 20 KIAS above my maneuvering speed, so my descent power is set accordingly. Think as if you are flying IFR here, not in a severe-clear, perfectly smooth day when you might want to run the speed up in the descent.

Example P210N cruise and descent power settings:

Phase	MP	RPM	TIT	Cowl Flaps
Cruise	26 inches	2,500	1,500-1,525°	Open 3 clicks
Descent	23 inches	2,500	1,500°	Closed

Again, I reduce power moving left to right: MP, RPM (I leave the RPM at the cruise setting in the descent), mixture leaned when the power is reduced, and cowl flaps closed. Some pilots think immediately leaning the mixture after reducing the manifold pressure is a strange thing. I lean to keep my cylinders warm (preventing rapid cooling) and to reduce unnecessary fuel flow—why pour gas out the exhaust? Keep in mind, this is a turbocharged airplane, so the engine can develop sea-level horsepower up to about 18,600 feet on a standard day; the reduction of manifold pressure changes the fuel/air ratio, not the ambient air density as in a normally aspirated engine.

In a normally aspirated airplane, as I descend I would have to increase the mixture as the air becomes more dense and the manifold pressure increases independent of the throttle position, or I would have to reduce MP as it creeps up during the descent.

Why do I like to descend at no more than 20 KIAS above maneuvering speed? Because I have been unfortunate enough to hit some very bad turbulence in both VFR and IFR conditions over that past thirty plus years I have been flying. I don't like turbulence very much anyway, but I really dislike turbulence that can bend my airplane. Most of us fly these airplanes on autopilot in cruise, and on a several occasions I have unexpectedly encountered turbulence that was strong enough to cause the autopilot to disengage. Now the autopilot is supposed to disengage under certain conditions, but if I have a high airspeed and hit this kind of turbulence, I want a margin of safety above the "bend-the-airplane" numbers.

If I am descending at 20 KIAS above maneuvering speed, I can get rid of 20 KIAS almost instantly just by extending the landing gear! I don't have to do a massive power change or anything else. I just extend the gear and there I am at the "safe" speed. Here are some rules of thumb.

Rules of Thumb:

(1) Extend the landing gear in level flight without changing power or altitude. The aircraft will slow 20 KIAS.

(2) Extend the landing gear in level flight and allow the aircraft to descend at the same power setting. The aircraft will descend 500 FPM and the airspeed will remain the same as it was in level flight.

(3) Reduce the manifold pressure by 5 inches and maintain the same pitch attitude and configuration. The aircraft will slow 20 KIAS.

(4) In normal flight operations, each 10 degrees of flaps lowered will slow the airplane 10 KIAS if pitch, power, and configuration remain the same.

Logic that flows from the rules of thumb above tells us:

(1) If in a climb, raising the landing gear will cause us to increase our climb rate by 500 FPM, if all other factors remain the same.

(2) If flying level, raising the landing gear will cause the speed to increase by 20 KIAS.

(3) Increasing manifold pressure by 5 inches will cause the airplane to climb at 500 FPM if all other flight control parameters remain constant (and the nose will rise about 4°).

(4) Raising takeoff flaps after departure while maintaining constant power and pitch will cause the airspeed to increase by 10 KIAS.

Leveling Off from the Descent

Why do I need to address leveling off from a descent? Because there are a couple of simple things that I have observed working with the hundreds of Centurion pilots I have had the pleasure to fly with.

One, you *can* hurt the engine if you don't do it right! Remember that in the turbocharged aircraft I leaned the mixture in the descent? Well, if I level off and decide to add power, I have to remember that *right to left* flow—I must increase the mixture before I increase the RPM or *push on that throttle*! Otherwise, I can damage my engine by unintentionally running it too lean.

If everything works out the way I would like it to when I level off, I won't increase the power. I will allow the aircraft to slow to 120 KIAS. It should do that with my descent power setting, assuming I have the wheels up, as this will set me up for an instrument approach or a traffic pattern entry. (The flaps will be up because I don't use them at high speeds or during normal descents.)

A Cessna 210 landing

9

Landings

Normal, Precision Instrument Approach, Crosswind, Short Field, Soft/Rough, Balked/Go-Around

Normal Landing

Let's start with a normal landing in a VFR traffic pattern.

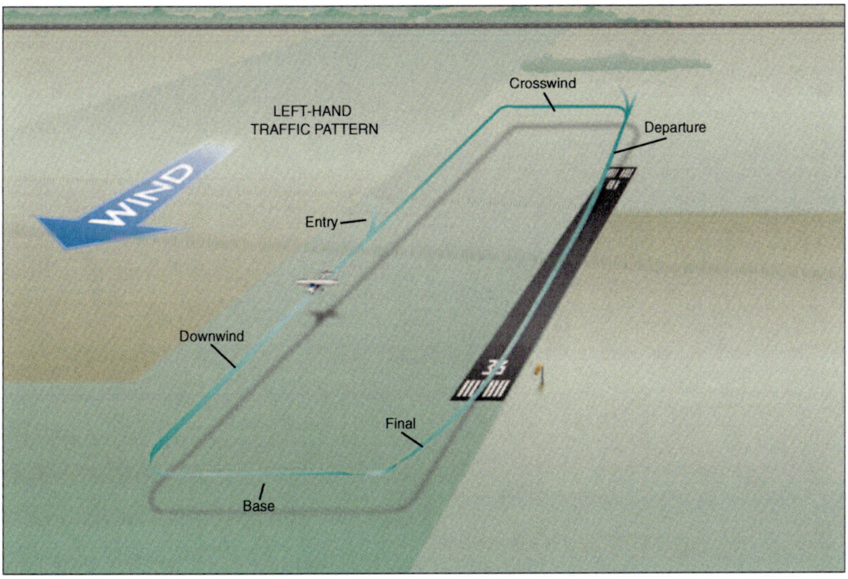

From the FAA's Airplane Flying Handbook

First, as you will find in Chapter 4 of the POH, the aircraft does not have an "approach flap" setting. Normal landings can be made with power on or power off with any flap setting desired. The aircraft can be landed with no flaps, partial flaps, or full flaps, and with or without power. It is comforting to know that we can land the airplane without flaps, because if we have an alternator failure, we may not be able to extend flaps depending upon remaining battery power, if any.

Cessna has published "Before Landing" airspeeds in ranges. As an example, these airspeeds for a T210L are 80-90 KIAS (flaps UP) and 70-80 KIAS (flaps DOWN).

In the "Before Landing" portion of the POH this is stated: "Wing Flaps — AS DESIRED (0° to 10° below 140 KIAS, 10° to 30° below 105 KIAS)," which is just a restatement of the allowable flap extension speeds from the Limitations section of the POH.

So, we can fly a traffic pattern and use no flaps to land, or we can use up to 30° of flaps in our landing process.

Landing gear stowed in the wheel well

But we *do* need the landing gear. If we have an electrical failure but still have hydraulic fluid, we can pump the gear down manually. Let's assume we can put the gear down in normal circumstances.

On the downwind leg we should fly the pattern per the FAA *Airplane Flying Handbook* and the Aeronautical Information Manual: position the aircraft parallel to the landing runway, one-quarter to one-half mile from the edge of the runway.

Somewhere in the downwind leg we must perform the pre-landing checklist. We need the gear extended, certainly before coming abeam the landing runway numbers. When we extend the gear, the nose will pull down and the tail of the aircraft will wiggle as the landing gear drops down, rotates the wheels parallel to the airplane, and moves them back up and forward into the locked position. This wiggle is more pronounced if the aircraft has gear doors, as they cycle open, deploy the landing gear, then cycle closed again.

You might note that when the landing gear first comes out you will need to feed in two or three strokes of left rudder trim and a couple of strokes of nose-up trim to keep the aircraft in trim.

When we are downwind with the gear down and no flaps, we need a +2 ½° nose attitude to maintain level flight.

In a normal landing situation where I have a runway of reasonable length (2,500 feet or longer) in the Cessna 210 and T210, I like to fly the pattern at 100 KIAS downwind, 90 KIAS on the base leg, and 80 KIAS on final. I use flaps as required. My personal method is I don't use flaps until I have the landing assured with power at idle, unless I am too high and need the additional drag to get me down.

Some pilots put the landing gear down and deploy 10° of flaps at the same time. When I've asked about their logic, I have heard everything from "that's what I was taught" and "that's what you are supposed to do," to "that way I don't have to re-trim the airplane," and a few other reasons. (There is nothing in the POH/AFM that says to use that technique.) I don't care for that technique for a couple of reasons. First, you are putting an extraordinary electrical load on the system by simultaneously using the two biggest power draws in the airplane. Second, I know that once I put those flaps down, I can't pull them back up if I have an engine failure without losing altitude. Since I have had a few engine-out events, I value the option.

In my typical VFR pattern, I fly 100 KIAS downwind with the gear down, flaps up, and the pre-landing checklist complete (including visually verifying in the gear mirrors that all three wheels are down and appear to be in their correct position with a green gear-down light). Power is set at about 20 inches of manifold pressure with that nose-up attitude. For the P210, typically 1 or 2 more inches of MP is required. For this airplane, I use a speed of 105-110 downwind based on landing weight, 95-100 on base, and 85-90 on final. The downwind power setting for the normally aspirated 210 is going to depend on the engine installed and the speed desired downwind, but is usually around 15 inches MP.

Abeam the landing runway numbers, I slowly ease the throttle to idle, lower the nose to maintain 100 KIAS, put in two strokes of nose-up trim to take the back pressure from the yoke off my hand, and put in two strokes of left rudder trim to center up the ball as well. Yes, I ease the power to idle if airport traffic will allow. I have already cooled my engine down during the descent, so I am not worried about shock cooling the engine.

Oops! No landing gear!

"The 210 Tuck"

Before we start a turn to the base leg of the pattern, I want to introduce you to what I call "The 210 Tuck." The Cessna 210 series has an interesting characteristic that causes a lot of pilots difficulty with their speed control when turning to base and final as well as during maneuvering flight. Our normal bank angle when turning base and final is around 30°. As you learned when earning your private pilot certificate, when you bank, the G-loading of the aircraft increases and you lose some of your vertical component of lift. As a result, the nose of the aircraft wants to drop—you expect that in most airplanes. Well, with the Centurion, as you bank through 20° the airplane starts "tucking" its nose down if you do not apply sufficient back pressure. The steeper you bank, the more the Centurion tucks its nose. Of course, as the nose goes down, the airspeed goes up. When we are in the pattern and turning base, we are normally trying to slow the speed, not let it remain constant and especially not let it increase. Watch for "The 210 Tuck" when you are out with your instructor for your initial training in the airplane.

Just before I start to turn base I look at my aim point and decide if I can deploy 10° of flaps. If so, I do that and keep the nose at the same pitch attitude that I had for the 100 KIAS speed. By deploying the flaps to 10° as I keep a constant pitch attitude, the aircraft will slow to 90 KIAS. As previously stated in one of those rules of thumb: all other things remaining constant, deploying 10° of flaps will slow the indicated airspeed by 10 knots.

Turning final is determined by my position relative to the aimpoint. I normally use what is referred to as the "Navy Pattern," i.e., I turn to final based on my position to the landing point. I don't try to make a square turn from base to final at a position extended straight out from the runway approximately ¼ mile or more. I then either raise the nose to slow to 80 KIAS or, if appropriate, I deploy more flaps and adjust the pitch to maintain 80 KIAS right into and through ground effect. I then commence my roundout about 10 feet above the ground and let the airplane fly itself into a stall just inches above the runway.

I then let the aircraft slow without applying brakes unless necessary.

Once clear of the runway and the aircraft is stopped, I utilize the "Clear of Runway" portion of the checklist.

I normally do not touch any controls or switches until I am clear of the runway and stopped. Those of you who have taken a commercial pilot checkride are well aware that in single-pilot operations, no ancillary controls are touched until clear of the runway and stopped.

Overcome the urge to unnecessarily switch the radio, raise the flaps, lean the mixture, or open cowl flaps before the aircraft is stopped. One of my students called me from Florida a few years ago and said, "Well, you were right—I raised the gear switch instead of the flap handle and the right main gear collapsed as I exited the high speed turnoff." This is not the positive reinforcement I was looking for concerning the procedures I teach, but it does point out the perils of doing too many things at the wrong time.

Precision Instrument Approach

Every pilot seems to have a different way they like to fly an approach, and some fly approaches with no particular procedure in mind. There are many ways an approach can be flown, just based on power settings, configurations, and desired airspeed. But regardless of how you fly it, a stabilized approach is best.

The technical definition of a stabilized approach can only be found in the ATP Practical Test Standard, so we aren't going to be reading it regularly. But let's say this: from the final approach fix (FAF) to the missed approach point (MAP), use constant power, a constant configuration, and a constant descent angle/descent rate (3° glide path is normal). This will yield an almost constant airspeed and will give us a stabilized precision approach. Once we reach the MAP, or before if we have weather above minimums, we can change power, deploy flaps, etc. as desired.

At some point we have to reduce the power to idle to land or we are going to fly the aircraft right into the runway at the instrument approach speed. You have to keep in mind that the stabilized approach turns into a landing at some point, determined by the pilot-in-command. From my perspective, the approach phase of the flight must end at the MAP, because when I reach the MAP I am either going missed or landing.

I like to fly a precision approach at a speed that is considered fairly fast by some pilots, with only the landing gear extended (this is not for a 2,500-foot runway, but one of 3,000 feet or more). With 17 inches of manifold pressure and the landing gear down, I am using an approach speed of 120 KIAS. I live in an area that can have some very low fog, and I always plan to go missed. When I see the runway before the MAP, I am happy I get to land; when appropriate I reduce power, deploy flaps as desired, and land. A key to this type of approach is keeping the 3° nose-down attitude into ground effect. Otherwise, if I haven't broken out of the weather when I reach the MAP, I merely increase the manifold pressure to the top of the green on the MP gauge while simultaneously raising the nose to +7° and commencing my climb. Then I clean up the airplane by raising the gear, ensuring the mixture is full rich, opening the cowl flaps, and putting in a whole bunch of right rudder trim. No pitch trim adjustment is required because I didn't have any flaps deployed. If you want to fly the approach at 100 KIAS, use 15 inches of MP as a nominal reference number until you know exactly which MP setting is right for your airplane.

At an approach speed at or below 90 KIAS, you are starting to fly the airplane below a speed from which you can easily execute a missed approach. If you are going to fly approaches at 90 KIAS or less, I suggest you and your instructor practice missed approaches regularly. If you have flaps down and execute a missed approach, you will find a very strong pitch-up moment and a strong left-turning tendency.

An Excerpt from the AOPA Review of 210 Accidents:

> Areas of Special Emphasis:
>
> The inadequate use of rudder has been sighted in numerous P/T-210 accidents, especially in balked landing and instrument go-arounds. The pitch-up and simultaneous left-turning tendency of the aircraft requires extra training over other aircraft. Stabilized

> and trimmed aircraft climbing straight ahead prior to turning is essential in the P/T-210 to preclude loss of control, especially IFR. The high number of accidents resulting from inadequate fuel shows a lack of knowledge and training in systems as well as ground servicing. All sampling drains must be identified and drained. Fueling must occur on level ground with the nose at the correct attitude. The tendency of many pilots to hold the aircraft on the ground too long before rotation is cited in several accidents, and excessive approach speeds count high in the accidents of P/T-210s running off the end of the runway because they are unable to stop on the available surface.

Since Cessna gives us a wide range of speeds and configurations in the POH/AFM to use for landing, you should find the right set of numbers and configuration that works for your airplane operating out of the fields you normally use. Once you find your numbers and get used to flying them concisely and constantly, you will find you can put the airplane right where you want just about every time.

Crosswind Landings

Your POH briefly addresses crosswind landings in Chapter 4. But there are no startling revelations about landing a C210 in crosswinds.

Excerpt from the POH:

> When landing in a strong crosswind, use the minimum flap setting required for the field length. Although the crab or combination method of drift correction may be used, *the wing-low method gives the best control* (emphasis added). After touchdown, hold a straight course with the steerable nosewheel and occasional braking if necessary.

I pretty much agree with that, except that I have never seen a crosswind in which I could land when I couldn't use 20° of flaps in the C210. It handles crosswinds very well if you apply the basic crosswind techniques you learned as a private pilot. Your POH may or may not give you a maximum demonstrated crosswind component. Some POHs have it, some don't.

A typical C210 demonstrated component is 21 knots. Just to refresh your memory, that refers to a 90° crosswind. That is a lot of crosswind! As you already know, but just to remind you, a demonstrated crosswind component is not a limitation. Landing in crosswinds is a pilot skill event. If you never practice it, you will never be good at it.

Using the wing-low, or sideslip, method of crosswind landing works well in a C210. You can also use the rudder trim to help with the rudder input. As an experienced mountain pilot, I prefer the wing-low method, because I want to know how much rudder I have left as I go into the flare to touchdown — and if I don't have enough rudder to safely land, I can go around. If you use the crab and kick method, you won't know whether you have enough rudder to keep the airplane tracking straight until after you have flared and are almost to touchdown. At this point, the aircraft will start to weathervane into the wind while you have full opposite rudder applied with your foot firmly on the floor and no rudder reserve.

Short-Field Landings

Here is what Cessna says in the POH for a 1979 T210N:

> For short-field landings, make a power approach at 74 KIAS with full flaps. After all approach obstacles are cleared, progressively reduce power. Maintain 74 KIAS approach speed by lowering the nose of the airplane. Touchdown should be made with the throttle closed, and on the main wheels first. Immediately after touchdown, lower the nose gear and apply heavy braking as required. For maximum brake effectiveness after all three wheels are on the ground, retract the flaps, hold full nose up elevator and apply maximum possible brake pressure without sliding the tires.
>
> At light operating weights, during ground roll with full flaps, hold the control wheel full back to ensure maximum weight on the mains wheels for braking. Under these conditions, full nose

down elevator (control wheel full forward) will raise the main wheels off the ground.

If you perform perfectly, the chart on the following page indicates what should be the results.

That procedure works, it's just difficult to get pilots to fly the Centurion that slowly on approach. Remember that you are essentially bringing the aircraft down to land at minimum controllable airspeed, just above stall speed. You have to be on your game if the landing area is really short. A couple of things can help:

- Practice, practice, practice. This is a spot landing at a low airspeed. You should practice this at the landing Weight and Balance you intend to use when you are really flying into that short field. You need to be able to "feel" the airplane and make many slight adjustments without thinking about them. Load the airplane to that projected arrival Weight and Balance and practice with your instructor the week before your trip.
- Aim point. You have to have one. It is the center of focus during this approach. Find a spot on the ground that you want to land just beyond. Fly to that spot.

A challenging approach over a mountain

LANDING DISTANCE

SHORT FIELD

CONDITIONS:
Flaps 30°
Power Off
Maximum Braking
Paved, Level, Dry Runway
Zero Wind

NOTES:
1. Short field technique as specified in Section 4.
2. Decrease distances 10% for each 10 knots headwind. For operation with tailwinds up to 10 knots, increase distances by 10% for each 2.5 knots.
3. For operation on a dry, grass runway, increase distances by 40% of the "ground roll" figure.

WEIGHT LBS	SPEED AT 50 FT KIAS	PRESS ALT FT	0°C		10°C		20°C		30°C		40°C	
			GRND ROLL	TOTAL TO CLEAR 50 FT OBS	GRND ROLL	TOTAL TO CLEAR 50 FT CBS	GRND ROLL	TOTAL TO CLEAR 50 FT OBS	GRND ROLL	TOTAL TO CLEAR 50 FT OBS	GRND ROLL	TOTAL TO CLEAR 50 FT OBS
3800	74	S.L.	725	1440	750	1480	780	1520	805	1560	830	1600
		1000	750	1480	780	1520	805	1560	835	1605	860	1645
		2000	780	1525	810	1565	835	1605	865	1650	895	1695
		3000	810	1565	840	1610	870	1660	900	1705	930	1750
		4000	840	1615	870	1660	900	1705	930	1750	965	1800
		5000	870	1660	905	1710	935	1755	965	1805	1000	1855
		6000	905	1710	940	1765	970	1810	1005	1860	1035	1910
		7000	940	1765	975	1815	1010	1870	1045	1920	1075	1970
		8000	975	1815	1010	1870	1050	1930	1085	1980	1120	2035

Figure 5-10. Landing Distance

- Pitch is airspeed — we all know that — we just need to work hard at it during this procedure. This approach typically requires *full nose-up trim* to reduce the force required to pull the yoke back far enough to fly at such a slow airspeed on final. A more aft CG (but within the envelope) helps with pitch control. Also, at this low airspeed, the airplane's nose is higher than during a normal approach and you need the seat as high up as you can get it so you can see your aim point. Be sure your seat is in the same position as when you practiced with your instructor.
- Think of what will happen if you need to go around with full nose-up trim on the airplane. There will be a strong pitch-up moment when the power is applied. You need to practice balked landings with your instructor and keep the process in the back of your mind during the approach (see "Balked Landings" below).
- The throttle should be closed at touchdown — that means pulled all the way back. Any residual power will significantly increase your ground roll.
- Use maximum braking with the yoke back consistent with the remaining runway. There is no need to stand on the brakes if there is plenty of room to stop, as might occur if you do indeed touch down right on the approach end of the landing area.

A couple of techniques I use depending on circumstances:

- This is a time when we really don't want our passengers chatting with us. Brief them ahead of time that you won't want them giving idle chatter during this approach; to speak only if providing important information, such as the location of other traffic or if the wheels aren't down, etc.
- The landing checklist is completed early, sometimes well before we are downwind.
- Be totally focused on the aim point and the airspeed. "Feel" the airspeed and sink rate.
- Keep power at idle even before touchdown — once in ground effect, if possible.

- Please be careful if you use this one — it is a departure from the norm with possible adverse results — as soon as I touch down, I raise my flaps up to dump any residual lift and put more weight on the mains for braking.
- Stop while going straight ahead. Don't let the airplane drift off center when applying maximum braking. You want equal weight on both wheels for the best effect.
- Execute an early go-around. If this approach isn't looking just like I want it to when I am 100 feet or more above the ground, I am going around. I am not going to let myself get down to the bottom of the flare and then decide I need to go-around. Do it early. If it takes a couple of attempts to be set up perfectly, take the time and do it right. Getting in to a tight landing area on the first try may be a desire, but it is unwise to attempt to turn a bad short-field approach into good landing. In some mountainous areas, the go-around may have to be executed as soon as you turn final and see you aren't where you need to be. A go-around, when appropriate, shows good aeronautical decision-making.

This mountain strip allows little room for error

Soft/Rough-Field Landings

Just as with this type of takeoff, Cessna does not address soft/rough-field landings in their POHs/AFMs. However, the FAA does, both in the *Airplane Flying Handbook* and in the *Practical Test Standards* for your private pilot certificate. If you use the technique set forth starting on page 8-19 of the *Airplane Flying Handbook,* you should do fine.

Gaining proficiency with soft/rough-field landings has its rewards!

The technique they describe is what I use on soft/rough fields in the Centurion. I put significant emphasis on keeping the weight off the nosewheel during rollout and while taxiing. As I have said before, this is a nose-heavy airplane and we don't want to drop the nosewheel into a soft spot.

Balked Landings/Go-Arounds

Cessna states:

> In a balked landing (go-around) climb, the wing flap setting should be reduced to 20° immediately after full power is applied. After all obstacles are cleared and a safe altitude and airspeed are obtained, the wing flaps should be retracted.

When you apply full power at a low airspeed with full flaps you are going to instantly have two very undesirable aerodynamic events occur in the Centurion: the nose is going to pitch up and the airplane is going to turn left — unless you prevent them from happening.

This retraction of the flaps to 20° immediately after application of full power is going to be a big help with the pitch forces, as reduction of flaps lower the nose in a high-wing airplane. But you are also going to need a lot of right rudder!

We know the airplane has these traits. You should have had them shown to you in your initial model-specific training and been taught how to deal with the situation.

Anticipate the event, be prepared to force the yoke forward while putting all the pressure required on the right rudder to *prevent* the nose from rising and turning left. Immediately retracting flaps to 20° helps with the pitch trim, but you will still need to quickly put in some down trim on the elevator trim and then a whole bunch of right rudder trim.

The key to success in the balked landing is airspeed. You must keep the nose low enough to maintain flying speed. At a speed below V_X the airplane may be flying, but it won't be climbing very rapidly, as the angle of attack is too high. *Push* the nose down to get the airspeed you want and trim the pressure off your hand.

Practice these skills until you can do them smoothly and correctly each and every time, and you will soon see that they aren't really that difficult. If you don't practice them, you won't be ready for what happens when you need to be.

Excerpt from NTSB Identification: DEN07FA165
Accident occurred Monday, September 24, 2007 in Moriarty, NM
Probable Cause Approval Date: 6/30/2008
Aircraft: Cessna T210M, registration: N732XE
Injuries: 1 Fatal, 1 Serious

The pilot trainee and instructor pilot were conducting the fifth flight in accordance with a training program developed by the operator. Satellite tracking data showed the airplane conducted four touch and go's, and satellite signal was lost on the fifth landing. A pilot-rated witness, who was located in his airplane at a taxiway/intersection, observed the accident airplane appeared to be set up for a short field landing based on the final approach descent angle and full flap position. Approximately 10 to 12 feet above ground level (agl), the airplane appeared to flare and then drop onto the runway. The airplane bounced and became airborne with the wings in a straight and level position. The airplane then drifted to the right of runway centerline, and the witness heard the "engine power come up." The airplane's nose suddenly pitched up and approximately 30 to 40 feet agl, the airplane stalled and impacted the terrain in a nose low attitude. The instructor pilot stated that he had no recall of the accident flight or the previous several days. Examination of the airframe showed the flaps were in the 30-degree extended position, and the elevator trim was approximately 10 degrees nose up. No anomalies were noted that would have precluded normal operations. The pilot's operating handbook supplement for "Balked Landing" procedures required the retraction of flaps to 20 degrees immediately. An elevator trim stall occurs when full power is applied to an airplane configured with excessive nose-up trim. Positive control of the airplane is not maintained resulting in a stall. These types of stalls usually occur during a go-around procedure from a normal landing. During the previous four flights, the instructor pilot noted on a grade sheet that the trainee was "below standards and needs additional training" on areas to include approach to landing stalls, short-field landings, and judgment. In addition, the instructor pilot noted the following comments, "[Trainee] sometimes struggles with aircraft trim, tries to 'muscle' aircraft...

Sometimes initiates flare too high, but has enough finesse for a safe landing — except for short field landings, where he doesn't have the airspeed to sacrifice. For short field approaches, needs to work on a stabilized approach and not flaring high."

The National Transportation Safety Board determines the probable cause(s) of this accident as follows:

The instructor pilot's failure to maintain aircraft control during aborted landing and attempted go-around which resulted in an inadvertent elevator trim stall. Contributing to the accident were the instructor pilot's inadequate supervision during the landing, improper recovery from a bounced landing, and the improper airplane configuration for the attempted go-around.

Execute a go-around here very early to avoid rising terrain

iPad mounted with a custom yoke clip

iPad mounted on an "ipadyokemount"

10

Cockpit Organization and Checklists

Everything has a specific place in the cockpit

CRM, meaning Crew Resource Management, has been around for a long time in the airlines and in large corporate aircraft. When operating a sophisticated single like the Centurion, I like to change the acronym CRM to mean "Cockpit Resource Management." As single pilots, we are the crew. That leaves us with the need to use all available resources when flying the airplane. The organization of the cockpit is one of those resources.

Let's start with the simplest device in the cockpit: the pilot's seat. We can adjust both the height and the back angle as well. The height of the seat can be singularly important for us. Some of us are short, some of medium height, and some of us are very tall. Some pilots have long torsos and some have a shorter torso. Therefore, it is not your overall height that necessarily determines where the seat needs to be positioned for optimum forward visibility, but the height of your torso. The Centurion has a high glareshield, so if you want to be able to use the cowling to assist you in your landing process, you need to have the seat up high so that you can see over the glareshield. If you have a very short torso, you might need to have a portable seat cushion and have the seat cranked up as high as it will go. My experience is that the higher you sit in the cockpit, the easier it is to land the airplane. That being said, sometimes you are so high that it is difficult to see some of the instruments when flying IFR. Some pilots I know have a seat height for VFR flying and lower the seat a couple of turns for IFR flying. Personally, I am able to use the same seat height for VFR and IFR.

So, you need a consistent seat position that you use every time you fly. If you are the sole owner/pilot of the airplane, that shouldn't be a problem. But if you are just one of the pilots that uses the airplane, it is advisable to verify the pilot seat height as part of your pre-flight process. Otherwise, you won't have the same sight picture you had the last time you flew (the angle of the seat back also changes your viewing angle of the instruments and the cowling). If you are flying instruments, the height of the seat changes your angle of view on the attitude indicator and the degrees of pitch you see on the pipit.

The rudder pedals are not adjustable, so if you must have the seat all the way forward to use the rudder and brakes but find it uncomfortable, you may want to consider aftermarket rudder pedal extensions.

Now that we have you seated, let's look around. Where are you going to put all these checklists, charts, chart binders, navigation logs, EFBs, IAPs, the POH/AFM, passenger briefing cards, the control wheel lock, the pulse oximeter, tablets, pencils, etc? Let's just make a simple list of things that can aid us:

- A kneeboard
- Yoke clips
- Pockets in the upholstery
- A center console
- A cockpit caddy on the floor
- A plastic tray under the seat
- Velcro on the yoke, side panel, and dash
- An EFB yoke mount
- Seat pockets on the back of the seats (virtually every 210 has them)

It is a commonly accepted philosophy in aviation that we should always have the same items in the same location in the cockpit so that not only do we always know exactly where they are, but we can find them in the dark should that be necessary.

Kneeboards come in all kinds of formats. Some are just a single board with a clip, but there are folding kneeboards, kneeboards with pockets, etc. I think the kneeboard you need depends not only on personal preference, but on the other aids in the cockpit. As example, see this center console below:

This console holds just about everything you need for the flight, except the kneeboard pad that you will be writing clearances on. However, if you don't have or want one of these consoles, then perhaps a different kneeboard would be appropriate — perhaps one that holds your IAPs.

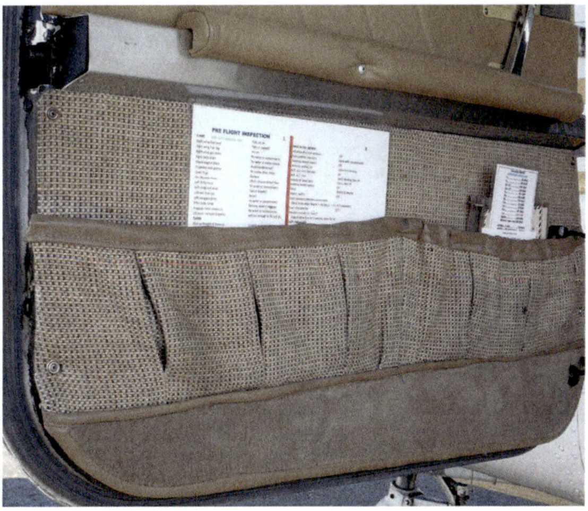

One aircraft owner uses a center console like the one shown above, but also puts selected checklists in door pockets:

Note that he has his IFR checklist and his Emergency Procedures checklists close by his left knee and easy to reach.

Another pilot has purchased a custom-made console, complete with armrest:

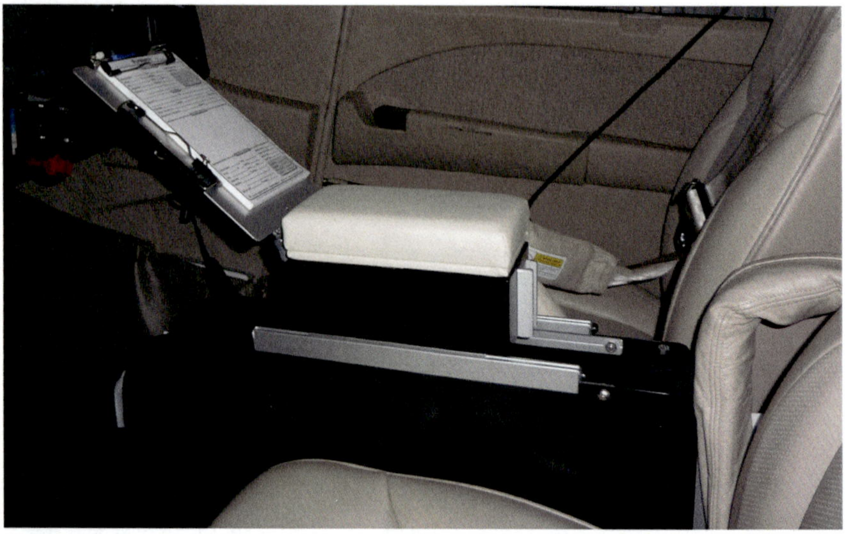

The "cockpit caddy" below was made by a pilot that flies VFR most of the time and he finds this serves his needs:

The iPad has brought the "EFB" into the general aviation cockpit. Most pilots I know have an iPad and use it for flight planning, weather, and in many cases for displaying their IAPs in the cockpit. The challenge is when and where is it mounted. Again, individual preferences should prevail, but the key is that it be mounted in such a manner as to be helpful, not an impediment in flight. My iPad is sometimes mounted to my yoke clip, such as when using it to display the IAP, and otherwise relegated to the center console for storage.

The use of yoke clips frees up your hands for other tasks and keeps charts off the floor and up where you can read them:

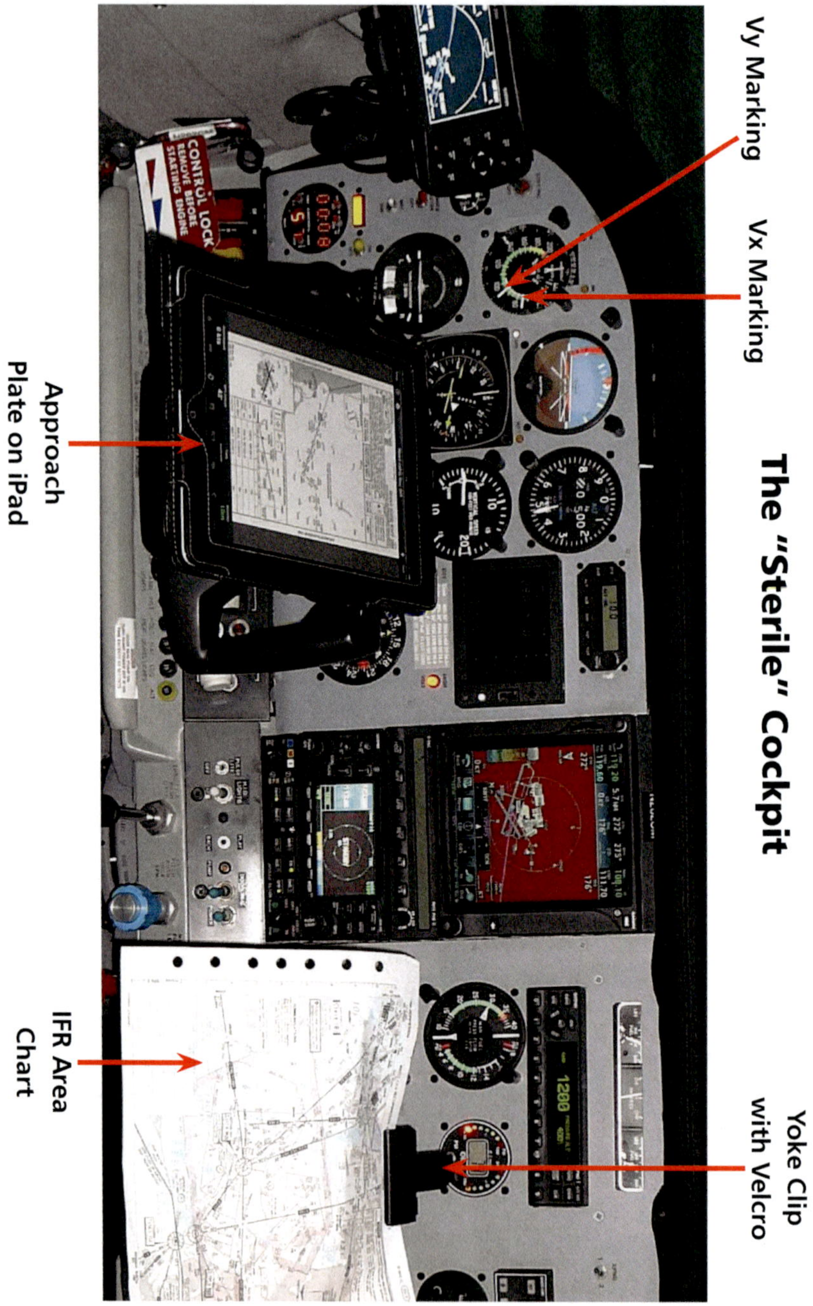

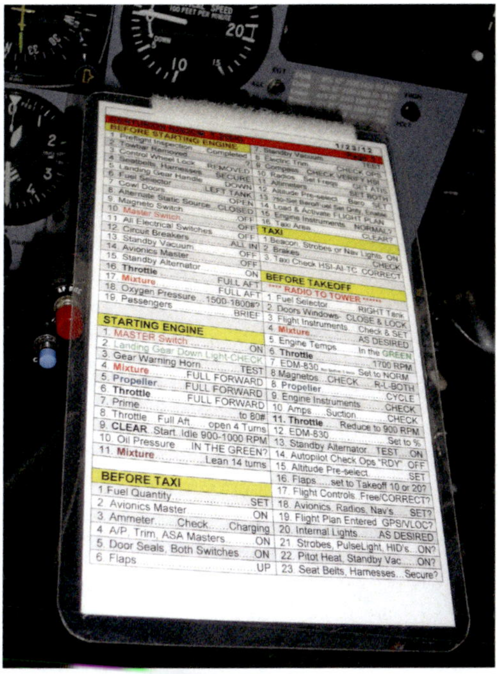

Checklists can be mounted and stored just about anywhere with the use of Velcro-like material.

The important thing is that they can be displayed when needed and stowed when they are not required.

On the following pages, you will see examples of checklists that may assist you in creating your own.

The more organized we are in the cockpit, the easier it is to fly!

Custom yoke clip

CENTURION N6586W P210/G

BEFORE STARTING ENGINE

1.Preflight Inspection...... Complete
2.Towbar and Chocks...REMOVED
3.Control Wheel Lock..... REMOVE
4.Pilots Door Window...........OPEN
5.Seatbelts, Harnesses ...SECURE
6.Cabin Doors.................SECURE
7.Brakes ....TEST
8.Landing Gear Handle ...... DOWN
10.Fuel Selector ....BOTH
11.Rudder Trim .... SET "T"
12.Pitch Trim ....SET
13.Cowl Doors ....OPEN
14.TKS Switch ....OFF
15.Mixture..................... FULL AFT
16.Prop .... FULL FORWARD
17.Throttle .... FULL AFT
18.Circuit Breakers ....ALL IN
19.Avionics Master ....OFF
20.Dump Handle........................??
21.Master Switch ....OFF
22.Magneto Switch ....OFF
23. Alt Static Source .... CLOSED
24.Pressurization Switch ....??
25.Cabin Pressure Dial.... +2K SET
26.Electrical (white switches) ALL OFF
27.Oxygen Backup .... Available?
28.Passengers.................... BRIEF

STARTING ENGINE

1.Master Switch .... ON
2.Landing Gear Down Light..Check
3.Landing Gear Warning Horn .test
4.EDM................ Set Fuel Quantity
5.WARNING LIGHTS ....Check 2
6.G500 ....ENTER
7.Mixture........... FULL FORWARD
8.Throttle .... FULL FORWARD
9.Prime (both switches)to 8 #, .OFF
10.Throttle-full aft –then... Open 1/4
11.CLEAR.Start.Idle 900-1000
12.Oil Pressure..... IN THE GREEN
13.Mixture.............. Lean - 8 Turns

BEFORE TAXI

1.Avionics Masters .... ON
2. A/P, TRIM,TIS Switches .... ON
3.Ammeter .... Charging ?
4.Flaps .... UP

FOR INFORMATION ONLY
CONSULT AIRCRAFT AFM

PAGE 1

5.Electric Trim.......................Check
6.Compass ....Check vs HSI
7.#2 Attitude Indicator...........CAGE
7.Engine Instruments...... NORMAL
8.GTN650 .... GET ATIS
9.G500 + #2 Altimeter...SET BARO
10.G500 set: .... HDG & Altitude
11.GTN650 set to ....Ground
12.GTN750 .... Enter Flight Plan
15.Transponder .... SET CODE
16.EDM 930............NO ALARMS ?

TAXI*** YOKE BACK to Taxi!!!!

1.Strobes? Nav Lights ? .... ON
3.Brakes............................ CHECK
4.HSI, turn coordinator....... CHECK
5.Attitude Indicators .... CHECK

BEFORE TAKEOFF – RUNUP

1.Radio .... to TOWER
2.Fuel Selector.....................BOTH
3.Flight Instruments...Check.... SET
4. EDM CHT's/Oil ...>200 each Oil >100
5.Throttle....................... 1700 RPM
6.Set EDM to NRM (3s LF button)
7.Magnetos...CHECK ....R-L-BOTH
8.Propeller ....CYCLE
9.Engine Instruments......... CHECK
11.Throttle.... Reduce to 1000 RPM
12.EDM....... set to %(3s LF button)
13.Autopilot.....(RDY)..... Check Ops
14.Rudder Trim......"T"............SET
15.Elevator Trim ....SET
16.Flaps ....SET 10
17.Flight Controls Free .... Correct ?
18.Seat Belts, Harnesses ... Secure
19.Avionics...radios........ navs set ?
20.Internal Lights ....As desired
21.Strobes, Pitot Heat.......... ON ??
23.Cabin Door, window and
Emergency Exit.... Closed...Locked
24.Cabin Door Handle Safely Lock
(pulled out).................UNLOCKED
26.TKS...primed .... ON ?
27.Air Conditioner ....OFF
27.Transponder ....ON ALT

TURN PAGE OVER NOW

Vx 80 Vy 100 Vglide 85

TAKEOFF
1.Fuel Pump Switches............OFF
2.Oil Pressure....... IN THE GREEN
3.Flaps....................................10°
5.Mixture...... 194+ pph **FULL RICH**
6.Power bot yellow33.5" 2700 RPM
7.Engine Instruments .. All Normal?
8.Lift Nose ....50-55 KIAS
9.Climb Speed.....Vy 100-120 KIAS
10.Landing Gear....................... UP
11.Flaps...above 85 knots......... UP

CLIMB (NORMAL) >1,000 AGL
1.Power .... 30" MP 2500 RPM -3
2.Mixture..................... FULL RICH
3.Airspeed ....100-120 KIAS
4.VERIFY.... GEAR UP, FLAPS UP
5.Cabin Door Lock.................. SET
6.Dump valve .... ?? CHECK
7.Pressurization........... ?? CHECK
8.Max diff pressure........... CHECK
9.?? Autopilot............................. ??

CRUISE Va 120 Vgld 85
1.Airspeed ....accelerate to cruise
2.Power..75%...26"........ 2500 RPM
3.Mixture...............................LEAN
-8 turns? .... 1500-1550 TIT
4.EDM CHT's 380-400
5.Cowl Doors ....CLOSE?
6.Rudder Trim .... ADJUST
7.TKS functioning properly .... ??

DESCENT
1.Enroute.........63% 23" 2500 RPM
2.63% power............... EGT <1500
3.TKS .... As Required

BEFORE LANDING...GUMP
1. Lighting .... As Desired
2.Seat Belts/Harnesses... SECURE
3.Fuel....................... AS DESIRED
4.**LANDING GEAR........... DOWN!!**
5.Mixture........ AS DESIRED-Lean?
6.Low Boost Pump .... OFF
7.Prop .... FULL FORWARD

N6586W Checklist 8 17 2011
Riley Intercooler Equipped -1" MP per 15°C

FOR INFORMATION ONLY
CONSULT AIRCRAFT AFM

Page 2

GAS – FUEL SELECTOR BOTH
UNDERCARRIAGE -
GREEN GEAR DOWN LIGHT?
MIXTURE AS DESIRED
PROP FULL (SCREW IN)

CHECK GEAR IN MIRRORS!

<u>GEAR DOWN? LOOK!**</u>**

LANDING (normal)
1.Check CABIN Pressure..... ZERO
2.Door Safety Lock.......... UNLOCK
3. Air conditioner...................... OFF
4.Airspeed 85

CLEAR OF RUNWAY
1.Throttle .... 1000 RPM
3.Mixture................... LEAN – 8-10
4.Flaps...................................... UP
5.Cowl Doors .... OPEN
6.Rudder Trim .... SET........."T"
7. Elevator ....Takeoff Setting
8.TKS...................................... OFF
9.Transponder .... VFR
10.Landing Light? Strobes?..... OFF
11.Radio ...... GROUND CONTROL

SHUTDOWN
1.Throttle .... 1000 RPM
2.A/C & Fan .... OFF
3.All Electrical (white switches)..... OFF
4.GTN FLIGHT TIME ..... RECORD
5.Avionics Master .... OFF
6.Mixture..idle cutoff ..... PULL OUT
7.Magneto Switch .... OFF
8.Master Switch .... OFF
9.Backup AI .... OFF

SECURE AIRCRAFT
1.Magneto Switch .... OFF
2.MASTER SWITCH .... OFF
4.Control Wheel Lock...... INSTALL
5.Pitot Cover..Sun screens........ON
6.Cabin-Baggage Door.... SECURE

FOR INFORMATION ONLY
ALWAYS REFER TO THE AIRCRAFT APPROVED FLIGHT MANUAL (AFM)

Centurion N6586W

OPERATING AIRSPEEDS

Vx	85KIAS
Vy	100 KIAS
Vfe 10°	160 KIAS
Vfe 20°	130 KIAS
Vfe 30°	115 KIAS
Vlo	165 KIAS
Vle	200 KIAS
Va	120 KIAS
Vso	58 KIAS
Vs1	73 KIAS
Vno	167 KIAS
Vne	200 KIAS
Vglide	85 KIAS
Window open max	200 KIAS

NORMAL CLIMB	120 KIAS
NORMAL VFR APPROACH	85 KIAS

POWER SETTINGS (- IC DIFF)

TAKE OFF..........35.5" MP 2700 RPM
CLIMB....................30" MP 2500 RPM
CRUISE..................26" MP 2500 RPM
DESCENT ....20" MP 2500 RPM
PATTERN 20" MP ....PROP FULL
INST APPROACH....17" ...PROP FULL

FUEL CAPACITY

TOTAL USABLE PER TANK:
FULL TANKS........................87 GALS
FUEL AT THE TABS ....64 GALS

OIL CAPACITY

TOTAL CAPACITY ....11 QTS
USE AEROSHELL.. SAE50,100WT AD

WEIGHT AND BALANCE

EMPTY WEIGHT ....2732 LBS
USEFUL LOAD ....1284 LBS
GROSS WEIGHT ....4000 LBS
MAX BAGGAGE WEIGHT..............200

Tire Pressures Nose-88 Mains-55

IFR CHECKLIST

BEFORE TAKEOFF

- Check both communications radios
- Check Marker Beacon lights
- Check both VOR's / GPS
- Check & Set – GPS/VOR CDI button
- Check Compass/HSI Heading Indicator
- Set Altimeters
- Check Alternate Static Source
- Check / Activate Pitot Heat
- During taxi, check for proper movement of gyro instruments and turn coordinator

APPROACH TO LANDING

- Obtain ATIS / AWOS
- Set COMM 2 to Ground Frequency
- Tune localizer to runway in use and verify
- Verify GPS or VHF approach Loaded and Activated
- Set OBS/Course Selector to Inbound Heading
- Check Marker Beacon Lights, Audio on
- Set NAV 2 as backup or set for missed approach and identify
- Set COMM 1 on Tower Frequency
- KNOW: 1. Initial Approach Altitude
  2. Decision Height (or MDA)
  3. First part of Missed Approach
- Check Heading indicator against compass
- Perform Pre-Landing Checklists (prior to FAF)
- **GEAR DOWN!!!**
- Acknowledge all Altitude and heading callouts
- Check Altitude remaining when within 500' of DH

MISSED APPROACH

- PITCH UP to appropriate climb altitude
- POWER UP to maximum allowable
- CLEAN UP – Set Gear, flaps, cowl doors IAW POH
- Verify Positive Rate of Climb
- Trim Aircraft
- Follow appropriate missed approach procedures
- Reconfirm avionics setup
- Recheck heading indicator with compass
- Review Climb Checklist
- Set up for next event

Cessna T210 at Biggin Hill Airport, London, England

11

Myths, Facts, Opinions, and Other Good Stuff!

The comments below are specifically directed to the Centurion: **FACT**

An exact copy of your POH/AFM in your office and/or at home is helpful: **FACT**

You can thoroughly preflight your 210 in ten minutes or less: **MYTH**

When fueling the 210, you can always count on the lineman to fill the tanks completely: **MYTH**

Using a properly calibrated fuel measuring tube during preflight could save your life: **FACT**

On warm days, there is no reason to warm up the engine: **MYTH**

The engine will warm up quicker if the cowl flaps are closed: **MYTH**

A cruise climb pitch attitude is typically 4°: **FACT**

When banking we will lose indicated airspeed. The amount of airspeed we lose is dependent on the bank angle: **PARTIALLY TRUE**

When banking we will lose indicated airspeed. The amount of airspeed we lose is dependent on the bank angle and the amount of time we spend in the bank: **FACT**

If we turn more than 90° while holding a 15° bank we will lose 5 KIAS: **FACT**

If we turn more than 90° while holding a 30° bank we will lose 10 KIAS: **FACT**

Semi-annual Recurrent Training is a good idea: **FACT**

If you want to slow down approximately 20 KIAS while remaining at your current altitude, extending the landing gear without changing power settings will do the job: **FACT**

A standard rate turn is a heading change of 3° per second: **FACT**

The bank angle for a standard rate turn is always 20°: **MYTH**

A large number of 210 accidents have been caused by running out of fuel: **FACT**

The person who flies the P210 infrequently can't "feel" the ball: **THAT IS MY OPINION**

When executing a missed approach or a go-around, the aircraft will have a tendency to pitch nose up and turn left, especially if flaps are deployed: **FACT**

When retracting or extending the landing gear, the wheels may extend as much as twenty-four inches lower during the cycling process than when the gear is in the down and locked position: **FACT**

RUDDER, RUDDER, RUDDER, then trim the pressure off your foot: **OPINION**

Before flying IFR in a GPS-equipped airplane, you have to check and ensure the navigation database is current: **FACT**

Have you ever gone to the airport just as you are leaving on a flight wondering (as the spouse and kids are hovering anxious to leave), have I got this all figured out? **I HAVE**

It takes about 20 minutes to fill the oxygen bottle(s) in a T210: **MYTH**

The lineman always knows where the oxygen filler port is located and has the correct tools on hand to fill your system without damaging the filler cap: **MYTH**

Gear mirrors are a good thing to have on the airplane, especially on the right wing: **OPINION**

On some of the 210 engines, I can run full manifold pressure in cruise for an unlimited time: **TRUE**

On most 210 engines, there is a five minute limitation on using full power: **TRUE**

Starting the engine and taking off immediately is good for the engine: **NOT TRUE**

If you are flying in cold regions, an engine heater could be helpful: **FACT**

Tire pressures never change when the aircraft is parked: **NOT TRUE**

If you are flying real IFR, a fully operational autopilot is valuable: **OPINION**

The approximate bank angle of a standard-rate turn can be estimated by taking 10% of the indicated airspeed and adding 5 (e.g., for 140 KIAS, 14 + 5 = 19° of bank): **TRUE**

Briefing passengers before a flight is not required: **NOT TRUE**

See FAR 91.107 and also FAR 91.519 for a truly "complete" passenger briefing

There is no need to calculate a Weight and Balance each time I fly: **DEPENDS**

If the landing gear won't go down electrically, I can always pump it down: **NOT TRUE**

There must be fluid in the system for the hand pump to work

If the gear pump continues to run for a minute or two after you have put the gear up or down, you should pull the gear pump circuit breaker: **TRUE**

See Section 7 of the POH

I do not need wing flaps to land a Centurion: **TRUE**

I can always get my wing flaps down to land: **FALSE**

The use of flaps for takeoff reduces the ground roll by approximately 30%: **TRUE**

An intercooler doesn't change engine operating procedures: **FALSE**

If I forget to lower the landing gear, a horn will go off before I touch down: **FALSE**

The gear warning horn is connected to the throttle, so it should go off when landing with power less than 15" MP, but…

You check the landing gear fluid by getting into the center console on the co-pilot side with a flashlight and a rag or paper towels in hand: **TRUE**

Tire pressures are the same on all models of the Centurion: **FALSE**

Chapter 7 of the POH/AFM, SYSTEMS, will show me the details of all the systems in my particular airplane: **TRUE**, *if* the aircraft has not been modified

The use of extra fuel tanks will not reduce my useful load: **NOT TRUE**

Air conditioning is a good thing to have in a P210: **OPINION, TRUE**

The Cessna 210 aircraft can only be operated from long paved runways: **MYTH**

The FAA has published more than 2,800 LPV approach procedures for use at nearly 1,400 airports since WAAS was enabled for operational use in 2003: **TRUE**

(as of August 2012)

When leveling off from a climb, I should lower the pitch attitude to level with the horizon as soon as I reach my desired altitude: **NOT TRUE**

Extending flaps in moderate to severe turbulence is a good idea: **NOT TRUE**

Extending the landing gear in heavy turbulence can help stabilize the aircraft: **TRUE**

You should weigh your cargo and passengers as part of your pre-flight: **DEPENDS**

When increasing power to climb from cruise, the cockpit flow of your right hand is from Right to Left: **TRUE**

(Cowl doors, mixture, prop, then throttle, then right rudder trim)

All passengers know how to connect their oxygen masks or cannulas: **MYTH**

You should brief your passengers before engine start on the hand signal you will use if you want them to stop talking: **TRUE**

If in addition to standard fuel tanks, your aircraft has tip tanks and a baggage tank—there could be as many as nine fuel drains to check! **TRUE**

We must modify the aircraft's Emergency Checklists to reflect any changes or modifications that have been made to the airplane: **TRUE**

Aviate,

Navigate,

Communicate!

Summary

The Centurion is an airplane that can do many things in the hands of a knowledgeable and proficient pilot. It can fly high, go fast, and haul six people and/or a lot of weight. It can operate in a wide variety of airport environments, from a dirt field in Australia to a gravel runway in Greenland to Los Angeles International Airport — VFR or IFR. It is comfortable, versatile, and stable.

From the accident reports included in this book, you have read about a variety of Centurion accidents and their causative factors. You will probably surmise that there are not a great number of different causes and types of problems pilots encounter in the airplane. However, the accident rate does seem to be higher than it should be. As you have read, it won't fly without fuel and it can be demanding of the pilot in situations such as a go-around or a missed approach. It seems like a lot of these accidents could have been prevented.

I believe the key to getting the most out of the Centurion is having it flown by a well-trained and proficient pilot. In my opinion, the Centurion is not unnecessarily demanding to the proficient pilot. But the operative phrase here is "proficient pilot."

I have had the pleasure of flying with well over one hundred different Centurion pilots, and they have had a wide range of aeronautical experience. Some of these pilot trainees were neophytes in the flying world, with under three hundred hours total time, while others had thousands of flight hours in a variety of different aircraft from the J-3 Cub to airliners and military fighters.

Perhaps because training pilots in these aircraft is what I do a great deal of the time, I have become particularly sensitive to things that I believe can cause grief when operating the Centurion. As with any high-performance airplane, you have to know your aircraft and its systems and have good situational awareness.

During my training programs, I strive to ensure the Centurion pilot is thoroughly knowledgeable of his aircraft systems and understands and is proficient in the integrated operation of those systems. The pilot must be able to detect when "something is not right," assess the situation, develop a corrective course of action, and execute it without getting behind the airplane. Good cockpit organization and the use of checklists will help every pilot, especially during times of emergencies.

No injuries, minimal damage (photo courtesy of Phil Ostroff)

The Centurion pilot must have the basic skills and capabilities that are needed to fly any aircraft of this type. He/she doesn't need to be the “Ace of the Base,” just well trained, competent, and proficient. Proficiency is the key.

We need three things to fly this airplane well:

- Knowledge of the aircraft and systems, attained through study and “hands on” training
- Skill, attained through training and practice
- Aeronautical judgment, attained through knowledge, training, and life experience

The insurance companies have found that model-specific training in the Centurion, particularly in the more complex pressurized Centurion, is critical to continued safe operation of the airplane. Many underwriters require initial model-specific training and annual recurrent training from an approved training provider. To the best of my knowledge, my Centurion training programs are approved by all U.S. underwriters that insure the Centurion.

The cost to insure a Centurion appears to be based on ratings held, flight experience, quality and frequency of training received, and age (of course accident history could also apply). Frequently, it appears that the cost of training is offset by lower insurance premiums, and some insurers specifically offer discounts upon completion of certain training programs.

I encourage you to train to a proficiency level consistent with the manner in which you operate your aircraft, including emergency procedures and considering the terrain over which you fly.

I believe recurrent training should be based on proficiency, or absence of proficiency, not just a period of time found on the calendar. If it has been five months since you flew actual IFR and you are about to launch on an IFR cross-country flight, you might be legal to fly, but are you proficient enough to handle the challenges that may occur during that flight? Spending an hour or two with your CFI before that trip could get you back up to the appropriate proficiency level needed to operate more safely.

Hopefully this book has been interesting and provided some helpful information to you. If you are already enjoying the pleasures of the Centurion, I trust you will continue to enjoy them for many more years.

If you have yet to acquire your 210, look long and hard — and get help from a knowledgeable person to assist you in the acquisition. During your purchase process, seek out a highly experienced 210 instructor to get you up to speed in your new aircraft and you should soon be experiencing the wonderful capabilities and pleasures of flying such a great airplane.

If you see me around the ramp one day, stop and say hi!

Fly Safely,

Chuck
Master CFI, CFII, MEI
www.SafeFlightintl.com

Appendix A

RAM Aircraft: Engine Shock Heating During Takeoff

The following is an article from the RAM Aircraft, LP web site: http://ramaircraft.com.

Conversations often center around "shock cooling."

So what is "Shock Heating?"

Shock heating is the situation that exists when a pilot almost instantly applies full power to the engines as he begins his takeoff. It comes as quite a shock to the engines. Reports suggest that the majority of engine failures take place during power changes — takeoff and landing. The pilot uses gear, flaps, speed brakes, gradual power reductions, and 1200° F to 1400° F EGT mixture settings to control shock cooling during rapid descents. RAM suggests the same degree of attention when applying takeoff power.

A rapid advance to takeoff power can be as hard, or harder, on the engines as shock cooling. Even at fast idle while the aircraft is taxied to the runway, the internal cylinder temperatures are only warm. Numerous engine components are not yet up to the heat and pressures of full power. The oil may be warm, but it's still cool enough to be at a higher than normal pressure. This situation sends misleading feedback pressures to the internal engine control systems. It takes time for the engine systems to perform their function and to adjust for the heat.

Bursting into full takeoff power can be compared to spilling an entire pot of hot coffee on your legs vs. one small cup. Although both have the same temperature, the full pot of coffee contains a tremendous quantity of heat. Damage results anytime anything is forced to absorb a large quantity of heat quickly—rather than gradually. RAM acceptance and test pilots use the following takeoff technique to help smooth their transition to higher cylinder temperatures and system pressures.

- Always assert your pilot in command authority to ensure the safe operation of your aircraft and passengers.
- When lined up on the runway centerline for takeoff, hold the brakes and smoothly advance the throttles to 27-30 inches of manifold pressure (MP). Allow the MP to stabilize 3-5 seconds. During that time, scan the engine instruments. Verify that they are normal.
- Release the brakes and smoothly advance the throttles to takeoff power manifold pressure. Remember, oil pressures and feedback signals to the controllers are adjusting as the temperatures rise. Pilots should allow engine systems time to function.
- Again, scan the instruments and make any necessary final throttle adjustments.
- Should an immediate takeoff be required upon arrival at the runway, the same procedures can be followed during the takeoff roll—if runway length is safe and suitable: partial power, stabilize, then full power.
- As you takeoff, cross checking fuel flow, RPM, MP and EGT is essential. It enables you to recognize and react to critical or premature instrument readings in a timely manner. Plus, it helps you remember the exact numbers when later communicating with your aircraft mechanic. Your aircraft mechanic needs to know specifics in order to effectively troubleshoot.
- Of further benefit, there should be no more surging and overboosting when using the above technique for smoother throttle advancement.
- Importantly, smooth power increases will allow the cylinder combustion chambers to experience a more constant and orderly transition to the higher temperature range of their design limits.

Appendix B

Type Certificate Data Sheet No. 3A21

Below is an extract to serve as an example of the information provided in a Certificate Data Sheet. This extract is for the P210N, and is found on pages 19 and 20 of the 45-page document that contains the Data Sheets for all Cessna 210 aircraft.

XV — Model P210N, Pressurized Centurion, 6 PCLM (Normal Category), Approved August 10, 1977

Engine Model P210N (S/N P21000001 through P21000760: Continental TSIO-520-P
Model P210N (S/N P21000761 and up): Continental TSIO-520-AF
*Fuel 100LL/100 minimum grade aviation gasoline
*Engine Limits Model P210N (S/N P21000001 through P21000760)
Takeoff (5 min.) at 2700 r.p.m., 36.5 in. Hg. mp. (310 hp.)
For all other operations 2600 r.p.m., 33.5 in. Hg. mp. (285 hp.)
Model P210N (S/N P21000761 and up)
Takeoff (5 min.) at 2700 r.p.m., 35.5 in. Hg. mp. (310 hp.)
For all other operations, 2600 r.p.m., 34.5 in. Hg. mp. (285 hp.)
Propeller and 1. (a) McCauley D3A34C402/90DFA-10
Propeller Limits Diameter: not over 80 in., not under 78.5 in.
Pitch settings at 30 in. sta.:
low 12.4°, high 28.5°
Model P210N (S/N P21000001 through P21000760)
Avoid continuous operation between 1850 and 2150 r.p.m. above 24 in. mp.
Model P210N (S/N P21000761 and up)
Avoid continuous operation between 1850 and 2150 r.p.m. above 23 in. mp.
(b) Cessna spinner 1250419
(c) McCauley hydraulic governor C290D4/T2
*Airspeed Limits (IAS) 1. Model P210N (S/N P21000001 through P21000150)
(See NOTE 4 on use of IAS) Never exceed 200 knots
Maximum structural cruising 167 knots
Maneuvering 130 knots

Flaps extended 115 knots
Landing gear operating speed 140 knots
Landing gear extended speed 200 knots
2. Model P210N (S/N P21000151 and up)
Never exceed 200 knots
Maximum structural cruising 167 knots
Maneuvering 130 knots
Flaps extended 115 knots
Landing gear operating speed 165 knots
Landing gear extended speed 200 knots
3A21 Page 20 of 45 Rev. 48
C.G. Range (Landing (+43.9) to (+52.0) at 4000 lb.
Gear Extended) (+42.5) to (+52.0) at 3800 lb.
(+37.0) to (+52.0) at 3000 lb. or less
Straight line variation between points given
Moment change due to retracting landing gear
(+3207 in.-lb.) S/N P21000001 through P21000150
(+2907 in.-lb.) S/N P21000151 and up
Empty Wt. C.G. Range None
*Maximum Weight 4000 lb. takeoff and flight
3800 lb. landing, 4016 lb. ramp, S/N 21000151 and up
No. of Seats 6 (2 at +34 to +46, 2 at +61 to +77, 2 at +101)
Maximum Baggage Reference weight and balance data
Fuel Capacity 90 gal. (89 gal. usable), S/N P21000001 through P21000760
90 gal. (87 gal. usable), S/N P21000761 and up
two 45.0 gal. tanks in wings at +43 (See NOTE 1 for data on unusable fuel.)
Oil Capacity 10 qt. (-12.5); 8 qt. usable
Control Surface Wing flaps Up 0° Down 30° +1°, -2°
Movements Ailerons Up 20° ±2° Down 15° ±2°
Elevator Up 23° ±1° Down 17° ±1°
Elevator tab Up 25° ±1° Down 10° ±1°
Rudder Right 24° ±1° Left 24° ±1° (measured parallel to 0.0 W.L.)
Rudder Right 27° 13' ±1° Left 27° 13' ±1° (measured perpendicular to hinge line)

For those who wish information on a particular aircraft, by serial number, the entire 45-page document can be found in a downloadable PDF file at www.FAA.gov.

Appendix C
Modification Suppliers

Gear Mirrors
Wilco, Wichita, KS

Anti-Ice Systems
Cessna Aircraft (booted anti-ice system)
Flight Ice, Inc. (TKS anti-ice system)

Autopilots
STEC
Bendix/King

Auxiliary Fuel Tanks
Flint Aero — wingtip tanks
O&N Aircraft Modifications — baggage compartment tanks
Monarch Development — older 210s only

Engines
RAM Aircraft, Waco, TX
Atlantic Aero, Greensboro, NC

Fuel Caps
Cessna umbrella style — Cessna Aircraft
Monarch Development replacement caps

Glass Panels
Garmin
Aspen

Intercoolers

Riley
Aircraftsman
Turboplus

iPad Mounts

Yuba-Sutter Aviation — www.ipadyokemounts.com
RAM

Landing Gear Door Conversion

Sierra Industries, Inc.

Standby Vacuum Systems

Aero Safe
Precise Flight

Standby Alternators

B&C Specialty Products, Inc.
Basic Aircraft Products, Inc.

STOL Modifications

Horton, Inc. (Horton STOLCraft)
Bush Conversions, Inc
Sierra Industries (R/STOL)

Speed Brakes

Precise Flight, Inc

Wing Tips

RMD (lighting within the wingtips)
Horton, Inc (replacement wingtips)
Met-Co-Aire (C-210A, B, C, D, E, F only)

Appendix D
NTSB Reports

From the NTSB web site (www.ntsb.gov):

> The NTSB aviation accident database contains information from 1962 and later about civil aviation accidents and selected incidents within the United States, its territories and possessions, and in international waters. Generally, a **preliminary** report is available online within a few days of an accident. **Factual** information is added when available, and when the investigation is completed, the preliminary report is replaced with a **final** description of the accident and its probable cause. Full narrative descriptions may not be available for dates before 1993, cases under revision, or where NTSB did not have primary investigative responsibility.

The accident reports found in this book were selected as representative reports that could be of interest to the reader and that point to specific types of accidents or causes of accidents involving Cessna 210 aircraft.

All of the accident reports were downloaded from the NTSB accident database and are available from the NTSB website. Some of these reports have been abbreviated for various reasons. Should you wish to examine the NTSB reports in their entirety, they can be found at www.NTSB.gov/aviationquery.

The author has no known knowledge or relationship with any persons involved in any of the accident reports herein.

DEN99FA034

Accident occurred January 20, 1999 near Albuquerque, NM
Aircraft: Cessna P210N, registration: N79NL
Injuries: 3 Fatal

History of Flight

On January 20, 1999, at 1527 mountain standard time (MST), a Cessna P210N, N79NL, was destroyed when it broke up in flight and collided with terrain in the Sandia Mountain Wilderness Area, near Albuquerque, New Mexico. The instrument rated private pilot and two passengers were fatally injured. Visual meteorological conditions prevailed at the accident site, but the pilot reported being in instrument meteorological conditions. An IFR flight plan had been filed for the personal flight being conducted under Title 14 CFR Part 91. The flight originated in Scottsdale, Arizona (SDL), at 1243 Pacific standard time (PST), and was en route to Wichita, Kansas (lCT).

According to the Federal Aviation Administration, the pilot telephoned the Wichita, Kansas, Automated Flight Service Station (AFSS) at 0606 PST, and began filing an IFR flight plan. The cellular telephone connection was lost but reestablished a minute later, and the pilot finished filing his flight plan. He then asked for an "outlook" weather briefing. The pilot was told there was an area of high pressure along his route, and that a weak trough would be moving across Kansas, Oklahoma, and Texas. Across the northern half of Arizona, the forecast was for scattered clouds at 15,000 feet; across New Mexico, scattered to broken clouds at 12,000 to 14,000 feet, with a chance of an isolated snow shower across the mountains; across Kansas, scattered clouds around 15,000 feet, becoming broken to scattered cirrus by the afternoon. Winds aloft at FL (flight level) 230 over Arizona were forecast to be from 300 degrees at 70 to 75 knots; over Albuquerque, 300 degrees at 60 knots; over western Kansas, 250 degrees at 30 knots; at Wichita, 250 degrees at 40 knots. The briefing was terminated at 0611 PST.

At 0729 PST, the airplane was serviced with 38 gallons of 100LL fuel. According to Corporate Jets' fuel invoice, both wing tanks were filled.

At 1235, the pilot was issued his IFR clearance. At 1237, he was cleared to taxi to runway 21. He was cleared for takeoff at 1243.

The flight proceeded without incident until 1519:10, when the Albuquerque controller told the pilot, "Verify you're level at [flight level] two three zero, [I'm] showing you four hundred feet low." The pilot replied, "We've just figured out we've had a dual vacuum pump failure. We've lost both our vacuum pumps, so we're gonna need to look for someplace. We've got electric back up systems here, but we're having a little trouble holding altitude and everything." The pilot then asked, "What's the bottom of the cloud layer? We're IFR at this time" When told cloud bases in the area were "around 13,000 [feet agl]," the pilot advised he wanted to descend and "get below the bases." He was cleared to descend and maintain 14,000 feet.

At 1521:47, the controller asked the pilot if he wanted to "continue on to Wichita or [if he wanted to] land early [at Albuquerque]." The pilot answered, "If we can get down below the cloud layer, we'd like to get to Liberal, Kansas, if we could." (It was later learned that the airplane was scheduled to undergo an annual inspection the next day at a maintenance facility in Liberal, Kansas.) The controller acknowledged and instructed the pilot to contact a different air route traffic control center (ARTCC) sector. Before the pilot could do so, at 1524:35, he reported, "We're having trouble. I think we're in a spin." Fifteen seconds later the pilot reported, "We're back with you," and said he was at 18,000 feet.

Eleven witnesses, six of whom submitted written statements, reported seeing a wing separate from the airplane, then observed the airplane spiral vertically to the ground. Three witnesses were at the bottom of the Sandia Peak Aerial Tramway. Their attention was drawn to the sound of an airplane engine "racing, as if it was pulling out of a stall or dive." The airplane "was already in a pure vertical spin" when one of the three witnesses saw it emerge from the cloud base. "There was no spiral component to its attitude." After the airplane made a couple of revolutions, he saw a "puff" at one of the wing roots, followed by the wing separation. The airplane continued to spin "in the pure vertical… traveling approximately 300 knots straight down." He added that the engine sounded like it was at full throttle.

Two witnesses were at a home when they heard and saw the airplane. One witness said the engine sounded "stressed out." The other witness said the plane "spiraled clockwise, at a tremendous speed, vertically into the ground…The left wing was missing…."

Meteorological Information

At the time of the accident, SIGMET Whiskey 6, AIRMETs Sierra Update 4 and Zulu Update 4, and a PIREP were in effect. SIGMET Whiskey 6, which included New Mexico and was valid at the time of the accident, called for "moderate occasional severe turbulence below FL 180 due to strong low level and mid level winds over mountainous terrain." AIRMET Zulu Update 4, which included New Mexico and was valid at the time of the accident, called for "occasional moderate mixed icing in clouds and precipitation between freezing level at FL 2000. Freezing level 8,000 to 12,000 feet."

Wreckage and Impact Information

Access to the accident site was gained on January 21, 1999. Light snow had fallen during the previous 24 hours. The wreckage was spread out in a fan-shaped pattern on a heading of 155 degrees. The propeller blades were buried in the ground underneath the engine. Next to the engine was the instrument panel. Nearby lay the wing carry-through spar with the left portion lying to the left of the flight path. Just beyond were portions of the left wing. The fuselage, horizontal stabilizer, elevator, and vertical stabilizer lay nearby.

The right wing could not be accounted for and it was determined that it had separated from the airplane during the breakup sequence. It was later found by hikers and recovered by the salvage company on April 15, 1999.

NTSB Identification: WPR10LA196

Accident occurred Sunday, April 04, 2010 in Leavenworth, WA
Probable Cause Approval Date: 06/20/2011
Aircraft: Cessna P210N, registration: N3599P
Injuries: 1 Uninjured

The airplane was in a cruise flight configuration when the engine ceased operating without warning. The pilot made an emergency descent and the airplane collided with a fence while landing in a field. Disassembly of the engine revealed extensive mechanical damage to the number two cylinder combustion chamber and piston assembly. The exhaust valve seat insert was missing and fragments of the insert were later found in the oil sump. The exhaust valve head had separated from the valve stem. The intake valve head had approximately 50 percent of the material separated. The fracture surfaces on the exhaust valve, intake valve and exhaust valve seat were damaged such that the mode of fracture could not be definitively determined. Hardness testing of the components indicted that they were within the manufacturer's prescribed specifications. The sequence of fractures and separations of components could not determined.

The NTSB determines the probable cause(s) of this accident as follows:

The failure of the number two cylinder exhaust and intake valves for undetermined reasons, which resulted in a loss of engine power.

NSTB Factual Report Aviation, ID: CEN10FA521

Accident occurred 09/06/2010 in Caledonia, MI
Aircraft: Cessna P210N, registration: N4632A
Most Critical Injury: None

On September 6, 2010, about 0845 eastern daylight time, a Cessna P210N, N4632A, operated by Centurion Aviation LLC and piloted by a commercial pilot with a certificated flight instructor (CFI) on board, sustained substantial damage when it impacted terrain during a forced landing following an in-flight loss of engine power while maneuvering near Caledonia, Michigan. A post-impact ground fire occurred. The instructional flight was operating under Part 91 without a flight plan. Visual meteorological conditions prevailed at the time of the accident . Both the CFI and the commercial pilot reported no injuries. The local flight departed from the Gerald R. Ford International Airport (GRR), near Grand Rapids, Michigan, about 0800. The pilot reported that the purpose of the flight was to conduct an instrument proficiency check (IPC). The pilot stated that the airplane was fueled to full prior to departure. He indicated that during the IPC he was cleared by GRR air traffic controllers to climb to 4,000 feet above mean sea level and hold as published over a GRR navigational aid. The pilot said that while he was starting the hold's outbound leg, he heard "a large bang from the engine area." The pilot removed his view limiting devices and turned north toward GRR. The CFI advised GRR air traffic controllers that the flight had a problem and they were returning to GRR. The pilot and CFI realized the airplane had no engine power and they agreed "to do an off field landing." The pilot set up for an approach to an alfalfa field. The pilot stated that the terrain was rolling and the airplane impacted terrain at the base of a hill in the field where its nose gear separated. The airplane "careened sideways" into a corn field and came to rest in the corn field. Fuel leaked out of the left wing's tip tanks and ignited. The pilot turned the airplane's fuel selector to off. The CFI opened the emergency door and both pilots exited. The fire was extinguished by the local fire department.

NTSB Identification: CEN11LA402

Accident occurred Saturday, June 18, 2011 in Albia, IA
Probable Cause Approval Date: 06/14/2012
Aircraft: Cessna P210R, registration: N900SR
Injuries: 1 Serious

The pilot stated that the airplane was fully fueled before takeoff. While en route, he requested a destination change due to deteriorating weather along his route of flight. During the descent, about 15,000 feet mean sea level, the engine lost power. The pilot's efforts to restore engine power were not successful, and he diverted to the nearest airport. The airplane descended below the clouds about 1,000 feet above ground level; however, the airplane was not in a position to land on the runway. The airplane ultimately lost too much airspeed and altitude as the pilot attempted to configure and align the airplane for a landing on the runway. He subsequently landed in a grass area about 100 feet southwest of the runway, nearly perpendicular to the runway orientation.

A postaccident examination did not reveal any anomalies consistent with a preimpact failure or malfunction. No useable fuel was observed in the airplane fuel tanks at the time of the examination. The fuel tank caps were securely installed, and no evidence of fuel staining or siphoning was observed on the airframe. The pilot noted that the fuel quantity indications before the loss of engine power were near empty and about 10 gallons for the right and left tanks, respectively. He added that the fuel gauges had never worked properly. The pilot stated that he did not visually check the amount of fuel on board during the preflight inspection. He commented that normal airplane endurance was about 5 ½ hours and the accident flight was about 3 ½ hours in duration. The airplane was not in compliance with a Federal Aviation Administration Airworthiness Directive requiring a placard denoting that the fuel level must be rechecked 2 minutes after fueling to ensure full fuel tank capacity. In addition, the airplane Pilot's Operating Handbook required a visual inspection of the fuel level as part of the preflight inspection.

The NTSB determines the probable cause(s) of this accident as follows:

The pilot's inadequate preflight inspection, in which he did not detect that less fuel was on board the airplane than planned, which precipitated a loss of engine power due to fuel exhaustion during initial descent.

NTSB Identification: CEN11CA132

Accident occurred Wednesday, December 29, 2010 in Abilene, TX
Probable Cause Approval Date: 05/26/2011
Aircraft: Cessna T210L, registration: N210BH
Injuries: 1 Uninjured

The pilot departed on a 615-nautical mile cross-country flight that reportedly would take him about four hours and fifteen minutes to complete. He reported that he had about five and one-half hours of fuel on-board, and that he received a weather briefing before departing. The briefing indicated that he would encounter marginal visual flight conditions en route to his destination. During the flight, stronger than expected wind conditions were adding about 10 minutes to his flight, and about 50 miles from his destination, he started his descent from 9,500 feet with a solid cloud layer below him. He contacted the air traffic controller and received an instrument rules flight plan for an approach to his destination. He said that as he broke out from the clouds, it was too late to make a landing at the airport, and the engine lost power. He switched fuel tanks and the engine re-started. He said that the controller asked him if he had enough fuel to a make an alternate airport, located about 44 miles away, to which he responded that the right fuel tank showed one-third full. About 3 miles from the alternate airport the engine lost power again, this time the pilot could not get it restarted. He then elected to conduct a forced landing on a road. During the forced landing the airplane's horizontal stabilizer impacted a signpost, causing substantial damage to the airplane.

The NTSB determines the probable cause(s) of this accident as follows:

The pilot's improper fuel management, which resulted in a loss of engine power due to fuel exhaustion.

NTSB Identification: CEN09LA233

Accident occurred Sunday, March 29, 2009 in Georgetown, TX
Probable Cause Approval Date: 12/29/2009
Aircraft: Cessna P210N, registration: N210BF
Injuries: 1 Serious, 1 Minor

The pilot was preparing to land when the engine lost power. He attempted to perform a forced landing to a road west of the airport. During the forced landing, the airplane impacted trees and then the road resulting in substantial damage to the airplane. An examination of the airplane revealed that the left main fuel tank was empty; however, the right main fuel tank and tip tank contained fuel. According to a telephone conversation with the pilot, during his pre-landing preparation he noted that the left fuel tank was "low" and switched from the left fuel tank to the right fuel tank. Shortly thereafter he noted that the engine had lost power. He attempted to restart the engine by turning the auxiliary fuel pump on and "cranking" the engine. According to the emergency procedures section in the Pilot's Operating Handbook the correct in-flight engine restart procedure requires the pilot to maintain 85 knots indicated airspeed, check fuel quantity, switch the fuel selector to the fullest tank, mixture set to rich, auxiliary pump to the "on" position for three to five seconds with the throttle open, then off, ignition switch to both, and advance the throttle slowly. An examination of the airplane fuel system revealed no contaminants and no anomalies. An examination and test run of the engine revealed no anomalies. The engine data monitoring unit showed a loss of fuel flow prior to the accident.

The NTSB determines the probable cause(s) of this accident as follows:

A loss of engine power due to fuel starvation as a result of the pilot's inadequate fuel management. Contributing to the accident was the pilot's improper engine restart procedures.

NTSB Identification: SEA08LA079

Accident occurred Saturday, February 09, 2008 in Carefree, AZ
Probable Cause Approval Date: 04/30/2008
Aircraft: Cessna P210N, registration: N4796K
Injuries: 1 Uninjured

The pilot reported that during the landing roll, he thought he "had blown the left tire" followed by the airplane settling onto its left side a second or two later. The airplane veered to the left and exited the runway, subsequently coming to rest upright adjacent to the runway. Post accident examination of the landing gear revealed that the gear warning horn was inoperative due to a disconnected cable between the throttle switch and the warning horn. Further examination, to include functionality testing of the landing gear, revealed no additional anomalies that would have prevented the gear from extending to the down and locked position.

The NTSB detennines the probable cause(s) of this accident as follows:

The collapse of the landing gear during landing rollout for undetermined reasons.

NTSB Identification: ERAIOCA276

Accident occurred Friday, May 21, 2010 in Cambridge, NY
Probable Cause Approval Date: 04/12/2011
Aircraft: Cessna P210N, registration: N234JK
Injuries: 2 Uninjured

The pilot stated that he and his passenger were practicing short-field landings at multiple airports. After completing a full stop, he taxied back to the runway and configured the airplane for a short-field takeoff. About 500 feet into the takeoff roll, the airplane became airborne with a high angle of attack. The pilot added full power and the airplane abmptly yawed to the left. The pilot corrected with coordinated rudder and aileron. As the airplane accelerated in ground effect, the right main gear touched down in a recently plowed field adjacent to the runway. The airplane bounced three times and the pilot aborted the takeoff, touching down in the field. The nose gear separated on contact, resulting in the airplane nosing over and coming to rest inverted. The empennage and wings ofthe airplane incurred substantial damage. The pilot reported no preimpact mechanical abnonnalities with the airplane prior to the accident.

The NTSB determines the probable cause(s) of this accident as follows:

The pilot's loss of directional control during takeoff.

NTSB Identification: NYC08LA152

Accident occurred Sunday, April 06, 2008 in Louisville, KY
Probable Cause Approval Date: 03/23/2010
Aircraft: Cessna P210N, registration: N734N
Injuries: 1 Serious

At the end of a long cross country flight, while on a straight-in approach to the runway, the pilot was advised by the tower that there was another airplane in front of him, and to perform S-turns for spacing. Two witnesses then saw the airplane begin a nose-high left turn, then saw the left wing dip. One of the witnesses, a pilot waiting for departure, noted that the airplane was in the landing configuration with landing gear and flaps extended. It appeared that the airplane did not increase airspeed, yet the more it turned, the steeper the angle of bank became. The witness then saw the airplane enter a spin with the left wing low and the right wing high. The airplane made 1 ½ revolutions before disappearing below a tree line. Tower controller observations were similar to those of the witnesses. Post-crash examination of the airplane revealed no mechanical anomalies.

The NTSB determines the probable cause(s) of this accident as follows:

The pilot's failure to maintain airspeed while maneuvering, which resulted in an inadvertent stall/spin.

NTSB Identification: WPR10LA140

Accident occurred Thursday, February 18, 2010 in Aurora, OR
Probable Cause Approval Date: 01/07/2011
Aircraft: Cessna 210L, registration: N777FW
Injuries: 6 Uninjured

The pilot reported that, after departing for a local flight, he attempted to retract the landing gear but the landing gear did not retract. He verified the position by looking out of the window and noted that the landing gear appeared to be in the down-and-locked position. The green landing gear position light was not illuminated. Because he thought that the landing gear was in the down-and-locked position, the pilot decided not to use the manual gear extension procedure. During the landing, the right main landing gear collapsed. Post-accident examination and testing of the landing gear system showed that it operated normally and no mechanical anomalies were identified. The underlying cause of the intermittent operation of the landing gear system was not determined. Testing of the manual gear extension system showed that it functioned normally. Had the pilot used the system, he would have received a gear down-and-locked indication and landed normally.

The NTSB determines the probable cause(s) of this accident as follows:

The failure of the landing gear to properly function for undetermined reasons, resulting in the right main landing gear collapsing during landing. Contributing to the accident was the pilot's failure to follow the emergency gear extension procedure.

Appendix E

Prepare to Survive!

by Chuck McGill

The following article, as well as many others of interest to the pilot, can be found at the SafeFlight, International web site: http://safeflight.ws/articles.html.

Throughout your initial and recurrent training pilot training, you are taught and re-taught how to deal with aircraft emergencies. The Prime Directive is to get the aircraft on the ground as intact as possible with minimal damage to the occupants.

Now that you have the aircraft on the ground and stopped, how are you going to survive? How are you going to assist in being rescued in a minimal amount of time? Common sense, a few inexpensive and lightweight items, and current technology are key to survival and rescue.

When flying out for the weekend $100 hamburger, it is very easy to be far from civilization in a very short period of time, even if flying a J-3 Cub or an Aeronca Champ. Even though you were just flying over the hill to the next airport, were you prepared in the event something unplanned occurred?

Any aircraft departing on a short cross-country flight should have at least minimal survival equipment on board. A survival kit should suit the terrain of the flight and the number of persons on board. A survival kit need not be elaborate, large, or heavy, and a small backpack can be used to hold everything.

As you know, survivability requires Shelter, Water and Food. Rescue usually requires signaling.

Shelter

Simple and inexpensive things such as lawn and leaf bags can provide shelter in many ways, such as a makeshift rain suit, a sunshade, a ground cloth, and even a method for collecting and storing water. Always carry some type of jacket and head covering. Think about how complex or how simple you want to be with your shelter equipment, and pack appropriately. I think any survival kit should have at least a dozen large lawn and leaf bags in it.

Water

You should always carry water in your airplane. It is the simplest of survival items, and one without which the body cannot function. Most folks just carry bottled water, but you can also purchase water in sealed survival pouches to be left in the airplane for long periods of time. Water can be purified through an inexpensive survival straw, but that may not help if you are in the middle of the desert.

Food

You don't have to be elaborate with food items. You could purchase military MRE meals, or dehydrated food from a camping store. But, for the weekend flyer, perhaps a jar of peanut butter and a handful of Power Bars will fill the bill. Sufficient food takes little space and can be stored for long periods of time.

Medication

Be sure you carry the medications that might be needed for you or your passengers on every flight, in case you have unintended delays. Band-aids, aspirin, tape, and a few other medical items you deem appropriate are of great value in a survival situation, and you probably already have plenty in your home. Small medical kits are available online from a variety of sources.

Signaling

Technology makes this fairly easy. Your cell phone won't work everywhere, but take it with you. It might work when you need it. Be sure the cell phone is turned off during the flight so you don't wear the battery down. If you have a satellite phone, take it with you. A PLB (personal locator beacon) is a handy thing to have. There is now an inexpensive PLB available that fits in your shirt pocket! It sells for $300 or less, operates on 406 MHz, and contains and transmits your location using GPS technology. If you have a handheld GPS in the airplane, it will provide your coordinates if you can use your cell phone or handheld radio to talk to others. Remember that commercial airliners and military aircraft guard 121.5 at all times. And they fly high, so even if you are in a valley you may be able to transmit to them in the blind. Be sure you carry spare batteries for all your portable equipment. Something as simple as a shiny space blanket, available in every camping store for less than $5, makes an excellent signaling device.

Basic Items

A Swiss Army knife or Leatherman Tool, a strong LED flashlight, waterproof matches, trash bags, water, a roll of duct tape, and some nylon rope. A waterproof aircraft survival manual created by the U.S. Air Force is available online for less than $15 [1]. An excellent survival guide is available in a deck of card format [2]. It is small, waterproof and easy to carry. The value of these survival manuals is that they help you focus on the important issues at hand, as well as provide survival information you would not otherwise know.

You can easily and inexpensively put together a basic survival kit that weighs ten pounds or less. A trip to the local Wal-Mart camping department can just about fill up your survival kit.

Of course you need to check your survival kit at regular intervals and replace time limited items such as food and batteries. I go through mine each year when I remove it from the aircraft during the annual.

Fly Safely,
Chuck McGill
Master CFI, CFII, MEI
January 2010

Chuck McGill is a retired Marine Lieutenant Colonel and Vietnam veteran. He received extensive training in sea survival, jungle survival, desert survival and cold weather survival during his twenty-two years of active duty.

Article References:

1. http://www.cardinalgear.com/aircrew-survival-pamphlet.html

2. http://www.safetycentral.com/surdecofcar.html

Appendix F

Practical IFR Flying

An instructional CD-ROM Course by Chuck McGill

Enhance Performance, reduce workload, and increase your level of safety when flying single pilot IFR!

This CD provides a full hour of instruction on methods and techniques to make Single Pilot IFR flying simpler, easier, less stressful, and safer!

Find out helpful things your instructor never taught you! Learn how using current technology can reduce your risk in IFR flight!

Practical IFR Flying is available from the following online vendors:

Aircraft Spruce
www.aircraftspruce.com

Marv Golden Pilot Supplies
www.marvgolden.com

Pilotsbooks.com

Made in United States
Orlando, FL
04 November 2026

90201372R00119